Second Edition

GIS Cartography

A Guide to Effective Map Design

Second Edition

GIS Cartography

A Guide to Effective Map Design

Gretchen N. Peterson

CRC Press
Taylor & Francis Group
Boca Raton London New York

CRC Press is an imprint of the
Taylor & Francis Group, an **informa** business

CRC Press
Taylor & Francis Group
6000 Broken Sound Parkway NW, Suite 300
Boca Raton, FL 33487-2742

Printed on acid-free paper
Version Date: 20140213

International Standard Book Number-13: 978-1-4822-2067-4 (Hardback)

Library of Congress Cataloging-in-Publication Data

Peterson, Gretchen N.
GIS cartography : a guide to effective map design / Gretchen N. Peterson. -- Second edition.
pages cm
Includes bibliographical references and index.
ISBN 978-1-4822-2067-4 (hardcover : alk. paper)
1. Cartography. 2. Geographic information systems. I. Title.

GA105.3.P47 2014
526.0285--dc23 2014002514

Visit the Taylor & Francis Web site at
http://www.taylorandfrancis.com

and the CRC Press Web site at
http://www.crcpress.com

For My Mother

Contents

Preface

In the first edition of this book, I started with an explanation of how this book came to be.

A lot has changed in the GIS world since the first edition of this book was published, particularly with regard to cartography and software development and the intertwining of the two. Map design has hit a new high and will continue to soar as design-minded programmers work to continuously upgrade the cartographic capabilities of our software tools. With more people than ever getting absorbed in the stories that only maps can tell, the audience has widened considerably, which makes our work more prescient, interesting, and ubiquitous. The talks and workshops I've given since the first edition was published have alerted me to one particularly exciting trend: an increase in the number of arguments surrounding cartographic design. These manifest in disagreements over whether traditional color palettes or modern color palettes are more effective, which software provides the most capabilities, what skills are needed to be a superb cartographer, and what exactly is wrong with all those bad maps out there. This dialogue, to me, means that people are seeing the benefits of effective map design and are caring more and more about how to achieve it. This second edition—which provides 24 updated examples, 20 new examples, and 2 new chapters—should further the cause.

With ten years in the geographic information system (GIS) business, I have come to learn how to communicate effectively with bosses, clients, other scientists, and the public using maps as the printed or digital culmination of my analyses. I am close enough to the beginning of my career, when I was blindly going about printing maps with no clear purpose or sense of style (as many other entry-level GIS professionals still do), to remember what it was like to make maps without a guidebook. Since then I have submitted hundreds of slides, report graphics, posters, and letter-sized results maps to bosses, clients, conferences, and colleagues. I wished that a book like this had existed when I began my career, and it was through that recognized need that I began thinking about writing one myself. This book is the result of ten years of research and practice in map design. It aims to give you a big boost up the map-design learning curve that I so painstakingly climbed.

Acknowledgments

There is no way this effort would have been possible without generous support and help from many people along the way. Many thanks to Susan Murray for the excellent graphics in Figures 5.2 and 6.55, Erik Jacobson for the excellent graphics in Figure 8.4, Stephen DeGloria and William Philpot for information on Munsell colors, David Howes who came up with the name for the book, Matthew Carter for information on his Georgia and Verdana typefaces, and Christopher Walter for several creative map design ideas. Thank you also to Jonathan Kirsch, Taisuke Soda, Alice Acheson, and John Lombard for helping to get the book started. Robert Berwick added considerable expertise to the geology section. Thanks go to Edward Mac Gillavry, Eric Fischer, and Frank Broniewski for providing resources and alternative views for the zoom-level chapter and to Dane Springmeyer for the technical review. Thanks to Tobin Bradley for lending a quote from his blog.

Thanks also go to Jay Watson and everyone at the Hood Canal Coordinating Council for kind words and support. Etienne Ouellet introduced me to some great web resources. Peter Fisher gave me access to one of his articles. Anna and Kyle Welling kindly reviewed the paragraph on Chinese color traditions for accuracy. Karsten Vennemann was instrumental in fact checking the web section, lending his own insights to the section, and helping to find temperature data when seemingly none could be found. The people of the Lone GIS Professional group contributed many fabulous book-name ideas, almost all of which were better than the original title, barring the few that were too hilarious to consider. Nonie Castro shared with me a thorough document that she authored on transferring project data from GIS software to graphic software.

Thank you also to my family for editing, critical review, chess breaks, and ice cream runs.

Trademark Names

Google Earth™ is a trademark of Google Inc.

Microsoft® and Virtual Earth® are either registered trademarks or trademarks of Microsoft Corporation in the United States and/or other countries.

MapQuest® is a trademark of MapQuest, Inc.

Yahoo! is a registered trademark of Yahoo, Inc.

McDonald's is a registered trademark of the McDonald's Corporation.

Staples® and that was easy® are trademarks of Staples the Office Superstore, LLC.

Microsoft® is a registered trademark of Microsoft Corporation in the United States and/or other countries.

Google Maps™ is a registered trademark of Google Inc.

Munsell® is a registered trademark of Gretag-Macbeth Holding AG.

Adobe® is either a registered trademark or trademark of Adobe Systems Incorporated in the United States and/or other countries.

Adobe Illustrator® is a registered trademark of Adobe Systems Incorporated.

Photoshop® is either a registered trademark or trademark of Adobe Systems Incorporated in the United States and/or other countries.

TileMill® is a registered trademark of MapBox LLC.

1

Introduction

In the first edition of this book, the intention was to make cartography accessible. It no longer needed to be relegated to the realm of stuffy academicians who looked down their noses at the admittedly meager attempts of the students and new professionals in the field. The idea was that with a more encouraging peer-to-peer tone, a huge increase in cartographic quality could be realized along with a commensurate increase in geographical and thematic understanding by the general public, which would ultimately be the benefactor of these better maps. In the first edition, I predicted that designers would infiltrate the field of cartography in large numbers and thereby raise the design-bar for those of us who came to the field by way of analysis and programming. By all accounts, this came true not long after the book was first published. Now we have designers using application programming interfaces (APIs) and graphics programs to create amazing-looking stand-alone maps, infographics featuring maps, and interactive map explorations on digital devices of all kinds. The bar wasn't just raised; it was thrown through the roof!

> A community is emerging; one that recognizes the potential for combining design's capacity for clarity and function with the accelerated pace and scope of science. … This is about science that needs to be communicated and design that has a mission. This is the exceptional and the inspired. Welcome to an emerging aesthetic of the new millennium.
>
> **Seed, March/April 2008**[1]

The good news is that we all have, for the most part, equal access to the tools and information needed to reach the roof and beyond. The proliferation of open source software is making it possible for anyone, regardless of financial resources, to attempt map creation. The best place to start is to simply find a good map visualization that communicates information in a way that makes sense for what you want to do and then to emulate it. You learn as you go. However, there comes a time when you wish you had a more comprehensive knowledge of the techniques to make a great map, so that you are sure you haven't missed crucial steps to take your map from good to great. This book still aims to provide that for you in a friendly yet detailed manner. It isn't a scratch-the-surface book. It presents you with what you need to know to do your job well.

People often ask me what the major pitfalls are in designing maps. Why, they want to know, are some maps simply awful, and what can be done to make sure those awful maps don't contaminate the field at large? My thought is that there are two major categories of awful maps and their faults are different for each. The first is the map made by the beginner cartographer, the student or early professional. The problems with these maps tends to be with layout organization, lack of color contrast, and clutter. If this is the stage of career that you're in, simply spending extra time to make sure that the margin elements look as good as the map, that the figure-ground relationship is readily apparent, and that there aren't too many features competing at the same level of hierarchy will make a big difference. The second category is maps made by people who have mastered the layout, contrast, and clutter issues, but haven't yet optimized font styling, generalization, and color trends. Fonts and generalization can benefit simply from the application of time and effort—not just using the default fonts and not just using the original data but taking the time to optimize it for the map scale. Color trends require a certain level of attention to the newest map styles and the styles that are permeating the public psyche the most (and are thereby the most effective at disseminating information).

If all of those issues are addressed in a map, then the map has a fighting chance to make a difference. This ability to make a difference—whether it's within a small circle of important decision makers or whether it's with the world at large—is why we make maps. Accuracy and a concomitant mindfulness toward ethics should be first and foremost goals for the geoprofessional—not simply trying to be the first to make a cool visualization out of some great new dataset. New datasets and cool visualizations are of course the drivers behind a lot of what we do, and the sense of urgency to be the first to explore them is important, but never underestimate the wrath of the general public if those early explorations turn out to be bunk. Take risks, but take them wisely.

On Design Experience

Many of us enter the geospatial profession without so much as one hour of design instruction in our formal education. This book seeks to fill that void by providing a reference that can be thumbed through time and again as you create your maps. Included are plenty of illustrations and novel concepts to kick-start your pursuit of mapping excellence.

I did not have any cartography training when I got started in the geographic information system (GIS) field fourteen years ago, and felt completely thrown into the deep end with no life preserver to help me out. There were very few books on the subject, and those were either outdated or too deep in theory for someone who wanted real-life examples and lots of pictures! Without a guide, the maps I made were probably mediocre at best. At that time, it wasn't as much of a professional *faux pas* for me, as a GIS analyst, to be so uninformed about formal map-making technique. People were usually impressed with the mere fact that a regular analyst could trot out a map *at all*, and they had next to no preconceived ideas as to how the map should look (except the ever-present and often misplaced concern about color, on which everyone seemed to have an opinion). This has changed somewhat. Today, people are still impressed with those who can make maps. But people have also seen a lot of maps online and in the news media, which have built up some good and some bad notions about how they should function, what they should look like, and what they are for. It's therefore even more important today to be aware of standards and styles.

Within this book, you will find a great amount of information on not only the tried-and-true traditional techniques, but also on the latest design skills that can really enhance your map products. You will learn that it is up to you to decide whether or not a particular accepted standard is right for your map or whether you should try a unique approach. You will learn how to cultivate your inner creative genius so that you are able to innovate. And where you need additional detail, inspiration, and instruction, there are some helpful references through which you can further your study.

Constructive Criticism

You know that map you just made? It needs some help. Be prepared to hear this and be prepared not to take it personally. Constructive criticism from a peer group can take your work to a much higher level than if you design in a vacuum. Incorporating critique into the design process is time-consuming, but without putting in the time and effort, the work may not resonate, it may not tell a good story, and it may not convey the right information. GIS analysts have always been a disparate group of professionals with backgrounds ranging from the natural sciences, computer science, and planning, to a whole host of other fields, and it is no different now. This means that we haven't always had the necessary design background to make sure that analyses are properly explained, displayed, and marketed within our organizations, on the web, at conferences, in front of community groups, in the news media, on mobile devices, and in printed publications. Thus, criticism is even more important for our field than for strictly design-oriented fields, yet it is usually only in those fields that we see design critique. This needs to change.

Many excellent cartographers are great at giving constructive feedback. Seek them out and carefully consider their critiques. However, every once in a while you'll encounter an elitist. The elite feel there is only one way to do something: their way. It's true that there are better ways and worse ways to make a map, but no *one* way to make an excellent map. For example, in a workshop I taught last year, we spread out five different map posters in the room. All the maps showcased the same data but in different ways, as they were all created by different mapmakers. The workshop attendees were asked to stand by the one they liked the best. There were definitely some maps in the group that didn't measure up. Everyone agreed on those, as emphasized by the fact that nobody stood by them. Intriguingly, there were two maps that were of the same caliber design-wise and the students split themselves more or less evenly between those two. With these two maps it all came down to their color schemes. Some liked the bold, more "youthful" color scheme map while others liked the map with the more traditional, subtle palette. While a few of the students argued about one being much better than the other, it became apparent that this was more a matter of color taste than of an actual difference in map quality. Both maps were equally effective in communicating the information. When responding to criticism, revise only when that criticism is coming from a true desire to make the map better, rather than an elitist mindset.

What Is a Geoprofessional?

I use the term *geoprofessional* in this book to mean those professionals who use GIS on a daily basis to produce analyses, data, and maps for the benefit and use of others beyond the geoprofessional himself or herself. Professionals who fit this description have a responsibility to create truthful and informative maps that elevate the profession, and to keep making cartographic advances so that data can be transformed into wisdom.

Tick-Tock Goes the Clock

Most of the guidelines found in this book consist of small touches that, when put together, make your map into a professional product. But you must be aware that though each technique may take up only a small amount of time in itself, the total time spent on all of the "little" touches that make up an entire map will be significant, especially when you are first practicing them. So what do we do about the time that it takes to build a good-looking map? How do we convince ourselves that the time it takes to make a sophisticated map is worth the effort and financial expenditure? How do we convince our bosses and clients? If *cartographer* is not in your job title, then this can be difficult, although you have certainly already decided that, to some extent, the effort it takes is worth it or you wouldn't be reading this book.

One way to be assured that this is time well spent is to consider the alternative scenario. Put simply, a hastily assembled map usually ends up being a bad map. And that bad map may have to be redone. This can actually lead to an even bigger time sink than designing the map correctly from the start. How many times have you had to reprint or re-export only to have to reprint several times to fix the errors? A successful and experienced mapmaker knows that a well-designed map takes time and effort, trial and error, and attention to the latest trends in colors, fonts, and so on. Successful mapmakers know this because they've *been there, done that*. But you can know this too, even at the beginning of your career, by reading through the design literature, like this book, and paying close heed to the standards, conventions, and the myriad other design considerations from the get-go. Even the nerdiest analyst among us can produce a quality map if only the requisite time and effort are put forth.

Not convinced yet? Try a little motivation through recognition. This might mean entering your map into a map design contest. Even if you don't win, you will likely spend way more time on the entry than on your day-to-day maps. This extra time spent on that one contest entry will increase the quality of all the subsequent maps you make, even if you don't pick up another design book again. That's because the skills you learn in making that map will stick with you over time. Another way to get recognition is to keep your map products for your yearly evaluation day. Nothing is more persuasive to a boss of a job well done than a good visual, and what better visuals to make your case than a "before I spent time on map design" map and an "after I spent time on map design" map?

Once you are a believer, you have to get your superiors on-board as well. These people will often need more convincing because they have likely never tried to create a map on their own and therefore have no concept of how much effort may need to go into one. Because of that lack of hands-on experience, the person who is paying for your time can be hesitant to grant you more of it. So some strategy to combat this is in order. One strategy, tailored for the Reluctant Boss, consists of taking a map-making assignment and creating two map products for it: the first is in the same style (or lack of style) as always; the second is completed on your own time by revising the map using the principles in this book coupled with any other reference materials that can help you arrive at a stellar end product. This second map will, and must, look fabulous (if not, try again). Take both maps to the Reluctant Boss. Explain that the second map was done on your own time, and be especially careful to describe how this new map is better than the first (in case this is a particularly design-blind boss). Also point out how many hours it took to come up with that map, but emphasize that it conveys the information more accurately, effectively, and is, in the end, a better tool. Then make your case that you would like to create more maps like the second one and need the time to do it. Pointing out the benefits to your boss of a better map is always helpful as well: increased professionalism, credibility, accuracy, client satisfaction, and effective communication.

Furthermore, the argument can be made that once a nice-looking map is made, a lot of the other maps that you make can build upon that initial effort and thereby take less time. You may only need to revise your map style once a year or once every few years to reflect the latest trends. If every map you make contains decidedly different data or extents from the last, then this strategy will obviously be of little use. However, you could potentially make a case for spending the requisite time to make a good map for the more high-profile map outputs with which you are tasked, and if the boss insists, spending less time on the minor map outputs.

Why Good Design Matters

In this book, you will see me advocating for good design that is both appealing and communicates its points effectively. I have heard the argument that aesthetic design is not as important as communicative design. Indeed, some say that a design could be entirely devoid of any aesthetic quality and still be considered "good" design as long as it communicates effectively. This makes sense in some situations. I think one could fairly assert that the McDonald's golden arches signs are good design, even though they are not at all beautiful. The same goes for a whole host of fast-food restaurant signs, which tend to use colors that are totally jarring from opposite sides of the color wheel. Sure, they look gaudy, but they get their messages across: "Eat here! We are fast and vibrant just like the colors on our signs!" However, does this apply to GIS maps? No! If we choose dissonant colors, viewers will not be able to look at the map for longer than 10 seconds before their eyes start to blur. If we are lax about balancing the elements on a page, then we risk losing credibility as it will look sloppy and unprofessional. If we clutter the map layout with a bunch of extraneous lines and boxes, we risk overwhelming the map with garbage. In short, an aesthetically pleasing map is also going to be a map that communicates elegantly.

Audience

The reader of this book is assumed to have a basic understanding of geography and GIS principles. An example is in the soils section of Chapter 6, "Features," which delves into techniques on displaying soils information in the most appropriate way. The focus in that section is almost entirely on the ways in which soils data can be *displayed*. Indeed, soils data are very complex and it would be beyond the scope of this design book to describe soils data in detail. In general, the book cannot discuss every aspect of every type of data that exists and every layout possibility. Therefore, descriptions are confined to the aspects of those subjects that pertain to map design.

This book is primarily aimed at early career professionals; they have an understanding of GIS processes and procedures, but generally lack a firm foundation in map design. Students of GIS, who might lack both GIS knowledge and cartography skills, will find it particularly helpful to use this book in conjunction with other texts. Alternatively, students can reserve this text for studying after a course or two in GIS have been completed. Mid- to late-career professionals will obviously have plenty of GIS analytical and procedural concepts under their belts as well as some level of sophistication in map design, but there still is a lot of material in this text that can be of use for this group, especially when it comes to displaying unfamiliar data or gaining inspiration for new map design techniques.

How to Use This Book

After reading the introduction, I strongly recommend reading Chapter 2, "Creative Inspiration," before proceeding. It is a quick read and will inspire you to add that extra bit of oomph to your map. You can keep that in mind while you are reading through the rest of the book. The other chapters can be read in whatever sequence makes sense to you. Chapter 3, "Layouts," does not go into detail about the design of the map element itself, but rather focuses on the overall page design and, especially, what needs to be on it and what doesn't need to be on it. In that chapter, the map is really just one of many *elements* that needs to be considered, albeit one that is of primary importance of course. The term *element* is used to mean any category of object that can go onto a map page such as the title, map, scale bar, and so on. That chapter is followed by Chapter 4, "Fonts," and Chapter 5, "Color," both of which explore the fundamentals of those two subjects and are important knowledge areas for both layout design and map design. Chapter 6, "Features," deals with map design in particular relation to common map features such as roads, streams, geology, land use land cover, and so on. While not every feature type that could possibly be put on a map is represented in that chapter, the selection of feature types was purposefully chosen to contain enough variety such that most mapping techniques are presented. Chapter 7, "Static Maps," discusses some of the intricacies in and tactics used for several common output formats: slides, reports, and posters. Chapter 8, "Projections," provides an overview of projection science and how it relates to map design.

Finally, Chapter 9, "Zoom-Level Design," delves into the exciting new world of digital, interactive mapping with specific design considerations for multi-scale cartography.

In many parts of this book I urge you to try and come up with novel and creative solutions to your design needs, but there is considerably more space devoted to cartographic standards and conventions. The reason is that without a foundation in standards and conventions, you cannot be completely comfortable with your creative solutions. While ignorance does sometimes lead to novel solutions, you, as the cartographic designer, need to know the fundamentals in order to assess whether a creative solution is satisfactory or not. Remember that the goal for most standards in map design is to lead to a cohesive set of rules for the symbology of elements so that a viewer can easily and quickly gain insight from a map. Indeed, as Colin Ware states in his book *Visual Thinking for Design*, "the goal of information design must be to design displays so that visual queries are processed both rapidly and correctly for every important cognitive task the display is intended to support."[2]

Skipping the How-Tos to Get Straight to the Good Stuff

You might flip through the pages of this text and wonder where the how-tos are. You know, the "click on this," "use this menu," type of instruction with screen shots and everything. That would be all well and good if it weren't for two things. First, focusing on how to make the software do what you want it to do gets away from our ultimate goal of achieving a great map product. In other words, don't let technology get in the way of the design. Second, how-to guides are necessarily focused on a single software product or group of products. With many sophisticated GIS products available today in both the proprietary and open source realms—and with many of us using not just one but a combination of these—it wouldn't be prudent to focus on a single platform. By keeping it general, the widest possible audience can be reached. After all, the focus here is on map design, which will always contain classic elements but also continuously changes with design trends and advancing technological capabilities.

Relative Map Scales

Throughout this book, there are a lot of references to map scale. You must know the difference between large-scale maps and small-scale maps to read the text effectively. Even geoprofessionals get this mixed up sometimes. The easy way to think of it is that large-scale maps show things as big and

FIGURE 1.1
These maps illustrate the relative terms: small scale, medium scale, and large scale.

small-scale maps show things as small. While scales in GIS are exactly defined and represented on the map as a scale bar or ratio, general discussions about map scale often use the relative terms: large scale, medium scale, and small scale. While the exact equivalents of these terms are somewhat ambiguous in the literature, the guidelines shown in Figure 1.1 will give you a general idea of what they are.

Small scale: 1:250,000 and smaller

Medium scale: 1:50,000 to 1:250,000

Large scale: 1:50,000 and larger

Endnotes

1. Originally published in *Seed* 2, no. 15 (March/April 2008), http://www.seedmagazine.com. Included here by permission.
2. C. Ware, *Visual Thinking for Design*. Burlington, MA: Elsevier, 2008, p. 14.

2

Creative Inspiration

We are going to have to face it: map making is in many respects a creative process. Even if you know all of the standard practices that exist for maps, you will still have to deal with your unique data and mapping goals. And the only way to deal with those is to employ your creative skills. A lack of those skills could be extremely detrimental to your achievement of cartographic excellence. This brings us to a concept of duality in map design: you need knowledge of mapping standards and creative intelligence. Without the standards know-how, you risk everything from leaving out a bit of information that could have been useful, all the way to making grievous communications errors. Without creative skills, you could make maps that look nondescript, that don't adequately illustrate your unique data and their ramifications, that don't increase your professional capital, and that fail to leave a lasting impression. That is why this book, which consists largely of explanations of standards, is only part of the equation. That doesn't mean that we will neglect the creative aspect of map making, however. This chapter is all about helping you to improve your design skills through creative exercise, and in other chapters there are reminders to do things your unique way or suggestions for unusual mapping techniques. With those tools you should be able to increase your creative skills on your own and thereby improve your map products considerably.

You might wonder if, through this discussion on creativity, I am advocating the use of elaborate and gratuitously ornate graphics on a layout. Indeed not! (Unless you are making a map of where all the elaborate and gratuitously ornate maps reside in the world.) Creativity can mean omitting a graphic element that does not add value even though it is found on most maps (logos are a good example of this). It can mean venturing to do things in a new way when the old way just doesn't seem to work anymore. Creativity means being open to new possibilities and being willing to implement them, even in the face of resistance from the traditionalists. And through this openness you will create maps that more closely honor their unique data and message than any robot-like mapmaker out there could.

This book, along with all the great tools, books, and other reference materials that are listed here, gives you a great foundation in classic map-making fundamentals and new cartographic paradigms such as digital, interactive mapping; export file considerations; using graphic design software; and so on. But where, you might wonder, are you supposed to acquire skills in creativity without going back to school to get a degree in art? Is it even possible, after say the age of 18, to become creative if you haven't particularly seen yourself that way in the past? Let's address those two questions, starting with some words about how it is most certainly possible to cultivate creativity no matter one's age or level of entrenchment in the analytical side of life.

You Can Be Creative

I've always enjoyed math, solving problems, and of course, being a geographic information system (GIS) analyst. Many would think this means that I am an analytical thinker. This would be true. But just because I know I am more naturally inclined toward left-brain thinking doesn't mean I have neglected my right brain.[1] From taking drawing classes in college, to painting water colors in my spare time, to visiting art museums, I conscientiously try to improve my creative skills. And you know what? They do improve from these activities. I see things as less black and white and more as a continuum. I see that there are many ways to present information to a viewer or reader and many of them can effectively communicate and be beautiful at the same time. I try to translate my creative experiences into all aspects of the work I do, from designing applications, to making maps, to writing reports and giving presentations. By coupling the creative brain with the analytical brain, it is much easier to create truly superior products that leave positive lasting impressions than by using one's analytical brain alone.

> Every great advance in science has issued from a new audacity of the imagination.
>
> **John Dewey, philosopher**

If you are skeptical about being able to become more creative through deliberate art-centric activity, perhaps it would help to point out that a lot of research has been going on in this area over the past 30 or so years and the overarching conclusion has been that creativity can be learned! A seminal book on the subject called *Drawing on the Right Side of the Brain,* published first in 1979 by Betty Edwards, argued that you can learn to use the right side of your brain through conscientious attention to it.[2] Her method to teach drawing, specifically, is through exercises in observation and fundamental drawing techniques. The main take-home message for us is that it is indeed possible to become a good artist and designer even if you haven't started out that way, simply by looking, learning, and practicing. The concept of learning through experience was highlighted in a survey of creativity professionals that concluded that learning through experience is the most important criterion for obtaining creative achievement.[3] Creative talent, in other words, is not just an innate ability that you either have or don't have.[4]

How do we translate this knowledge—that we can become creative if we try—into practical methods for success? There are two simple, concrete methods to practice, loosely labeled *doing* and *seeing*. To explain these, think about a typical art class. Students in a typical art class are taught by being required to create their own art (doing) and by viewing examples of art (seeing).

Doing

The typical geoprofessional is already a step ahead in these methods of seeing and doing because we already practice the doing part as a natural component of our professional work. What I mean is that we are already making maps, sometimes on a daily basis, and therefore do not lack opportunities to try out techniques that we see and ideas that we form as a result of seeing. Therefore, it is not absolutely necessary to make an effort to practice creativity. Now don't think that lets you off the hook completely. Trying to stretch ourselves by creating other types of artistic end products other than maps can still be a highly educational endeavor. Some types of activities that would help are taking an art class, or keeping a sketch book to record what we see (try this instead of snapping pictures), or drawing a cartoon for a child's entertainment, and so on. However, as I said, these aren't absolute necessities for people in our profession if you just can't bring yourself to undertake alternative creative pursuits.

Seeing

Seeing can be defined as the process of actively and deliberately observing and absorbing the imaginative creations of others. Geoprofessionals are often lacking in this realm. Ask yourself how many times you've sat down and studied a book full of maps in order to gain inspiration for projects, for example. What are some ways to practice seeing? The easiest and probably most important way is to view art (not just maps) on a regular basis. You can accomplish this either in large chunks of time each year or on daily forays, but however you do it, rest assured that it will make a significant impact on the professionalism, communicative ability, and memorableness of your map designs. Visit local museums and gardens, view architecture, read any of the multitude of art and design books, and peruse magazines in any of these genres. You can most certainly include the study of maps from classic and contemporary cartographers in this list of seeing activities as well, but don't just limit yourself to just maps.

While it might be best to take yourself away from your element (the computer) to practice seeing, you can start by browsing exhibitions of outstanding art galleries, such as the Smithsonian American Art Museum, the Louvre, and the National Gallery of Art, or visiting the David Rumsey Historical Map Collection, all online. While there are many design-centric texts out there, my personal favorites are books by Edward Tufte, which illustrate how one can both follow some of the accepted design techniques while questioning others. And finally, as Garr Reynolds of the Presentation Zen blog so aptly puts it, "getting away from the computer, getting off the grid and finding time alone ... is crucial to keeping the creative spirit alive."[5]

An Example of How to See

I was at the Seattle Art Museum with my daughter and mother, not to seek design inspiration, but just to have a fun outing with the family. But I *still* came away with some ideas for maps and layouts even though I hadn't even intended to do so. One of the ideas came from my daughter who was six years old at the time. At that age, kids have a knack for noticing things that our adult minds have become so accustomed to that we fail to notice them at all. I was showing her some portraits since she was interested in how people recorded their likenesses before cameras were invented. My mother asked her why she thought the lady in one of the portraits was frowning and my daughter replied that the lady was frowning because her portrait was in a less-fancy frame than the one on the man's portrait next to it! Now, most of us wouldn't have even noticed the picture frames at all, or only in a cursory fashion. I took note of this and thought about how I could apply it in some creative way to the next map I was going to make. By the time we left the museum, I had made a mental note that if I ever needed to make a map of historical data, I ought to try and use a fancy gilded frame as the layout border.

If I hadn't been writing this essay on creativity when I went to the museum, I may have just chuckled at my daughter's insightful reply but not applied it to my work. That is why you should keep these creativity concepts in the back of your head when you are viewing art, so you are ready to file away any pertinent and applicable ideas to use back in the office. If you start practicing being open and aware of new ideas, you might feel flooded with hundreds of disparate ideas, all of which seem like good things, but which you have trouble organizing and putting to good use in your maps. At this stage you need to think about carrying a small notebook around with you so you can jot these things down. An alternative to the notebook is to use your cell phone to snap pictures of anything that strikes you, but only if you think the visual will be enough to remind you how you could apply it to your work.

A SURE-FIRE WAY TO GLEAN NOVEL IDEAS FOR YOUR MAPS—INVITE A NOVICE

Try this little trick the next time you are seeking a novel approach to solving a map design problem: invite a nongeoprofessional, perhaps a teenager, or just someone who doesn't have much experience with maps, to help you out. Their lack of preconceived ideas about mapping will allow them to see your problem with fresh eyes, and without any background mapping skills, all of the novice's ideas will be creative. There's nothing like a little ignorance to bring about some useful new inventions. Your job is to weed out the unusable ideas and see if you can get one or two good ones.

Applying All of This to Your Map

When you are open to new ideas, you can create a finely tailored map design that uses space wisely, emphasizes the uniqueness of your map data and subject material, and solves novel design conundrums. For example, a colleague of mine who practices the creative process quite often by both making maps on a near-constant basis as part of his work and keeps up with the art world through cultural excursions, recently came up with two great solutions to map issues he was having. For one, he wanted a slightly unique way to separate the many pieces of a margin grouping box. His solution was to provide titles for each section of the margin, but to both deemphasize the titles and use them as graphic elements by running them vertically along the separations between the margin sections. So, for example, between the legend and data sources sections he simply wrote "Data Sources" vertically in the space between them. Another example is his use of what I call the "split screen" approach, where he was faced with large amounts of data on either side of a national park that he wanted to display on his map. The national park itself, however, was not a central part of the map's purpose and did not contain any data that he needed to display. His solution was to cut out the national park and allude to it by presenting the map as two separate parts next to each other, separated only by a small amount of white space in between. An inset map showed the entire extent of the data to make this clear to the viewer.[6]

Getting started with a new map-making effort can be very daunting. Staring at that blank page, knowing you have umpteen elements that will have to be arranged on it in a meaningful and good-looking way, and not having a clue how you will do it is scary. People are telling you they want photographs on there, you know you need call-out boxes, and how are you going to pick a color scheme out of all the options available? Well, relax, because there's one easy way to get started when this happens.

Use a variation of a map that has already been made. To begin, start looking up other maps that you like. Keep a book of maps by your desk for such purposes, go online, or sift through other maps that your department has done in the past. There are two ways to produce derivatives of these other works. The first is to find a layout that you really like and try to follow its general design on your own map including placement of the title, map element, and so forth. The second is to compile a list of things that you like from each map that you want to apply to your own map such as color schemes, emphasis map placement, title fonts, and so forth. The list can be consulted when you put together your own design so that you wind up with an amalgam of other map designs on your own. Whichever method you use, you need to make sure that the final product isn't an exact replica of the original. However, this isn't hard to avoid because even if you begin with the idea of using an exact copy, you soon realize that all the unique qualities of your own data, audience, and communication needs will necessarily alter the finished product to such an extent that it no longer resembles the original. (Conversely, an inspiration map could actually lead you to *not* copy something that you find particularly distasteful!) Sometimes we are stifled by our own desire to start from scratch and thereby create a truly original work, thinking that this is the way that everyone does it. However, that isn't true, because almost all works of art are derivatives—related to the art that has come before them.

Some recent examples of maps that pile innovative ideas on top of traditional concepts include

- City maps of languages used in social media. These are point-based maps that employ a dot schema coupled with a color palette with each color denoting a different language. These techniques are traditional. However, the innovative cartographers recognized that their point data was so dense that they could get away with not having any background context. The background is all black and the dots are in bright hues. The spatial context is implied by the location of the dots themselves.[7]
- A digital, interactive commuting time map. This map is innovative in its use of an unusual hue: pink. While pink is actually a color used sparingly on many historic political maps, we don't often see it as the main color on modern maps. However, the color is refreshing since it is so unusual. The rest of the map is derivative in that it employs a typical (but effective) choropleth style with a single hue gradient.[8]
- A place name map. This political map shows place names in standard cartographic style, in a digital map format. The derivative part of the map design is that it shows standard cartographic place

names (e.g., New York, Vermont) just as with any other political map. The innovative part of the map design is that there is no basemap included for context, just the place names and a few place points. Without the basemap to distract the map reader, it definitely draws attention to the names.[9]

Another great way to get started with a map is discussed in Chapter 3, "Layouts," in the section titled "Emphasis Maps and Wireframes." An emphasis map, simply put, is just a simple sketch on paper that you make to help organize the layout's *big picture*. By arranging elements on paper, you are allowing yourself to make mistakes, to change things around when they don't look good, and to make many iterations until you find something that appeals to you and the specific mapping needs of your project. Start by looking at Appendix A, "Layout Sketches," because it contains several of these types of sketches, which will give you some organization ideas. Indeed, you can combine the two techniques—using an inspiration map and an emphasis map—by finding an inspiration map, sketching out the elements of the inspiration map, and then creating an emphasis map that is an amalgam of the inspiration map and your own map's special characteristics.

Summary and Final Prodding

While our daily activities almost always involve a lot of computer time doing data analysis, creating data, developing applications, and producing maps, we need to occasionally extract ourselves from this environment in order to refresh our creative skills. Feeding our inner creative genius so that it flourishes allows us to make a lasting impression with our maps. In so doing, we make a difference to our map audience and put our career trajectories on a higher plane.

Creative Maps

The following are some sources of creative maps to start your search:

Wind Map: http://hint.fm/wind/

Women's Political Rights around the World: http://777voting.com/

Susan Stockwell: Mind the Map: http://www.susanstockwell.co.uk/exhibitions.php

Katharine Harmon and Gayle Clemans, *The Map as Art: Contemporary Artists Explore Cartography* (Princeton, NJ: Princeton Architectural Press, 2010)

Frank Jacobs, *Strange Maps: An Atlas of Cartographic Curiosities* (New York: Viking Studio, 2009)

A Few Places to Start Seeing Art from Your Desktop

These are just a few of the multitude of online options for viewing art. Links to other exciting online art exhibitions are listed on the author's website: http://gretchenpeterson.com/links.php

Smithsonian American Art Museum: http://americanart.si.edu/collections/index.cfm

Louvre: http://www.louvre.fr/en

National Gallery of Art: http://www.nga.gov

David Rumsey Historical Map Collection: http://www.davidrumsey.com/

Museum of Modern Art: http://www.moma.org/

Endnotes

1. Purists would remind me that the term *right brain* is an overused and meaningless expression. According to Andrew Razeghi in *The Riddle: Where Ideas Come From and How to Have Better Ones* (San Francisco: Jossey-Bass, 2008), right-brain thinking is a myth: "Although the right hemisphere may generate more possibilities, it does not select one over another and therefore needs the left hemisphere. You use your entire brain in the creative process." The reason I continue to use the terms *right brain* and *left brain* is merely because they are easily understood terms that distinguish between analytical thinking and creative thinking.
2. The book has been updated and is now called *The New Drawing on the Right Side of the Brain*, by Betty Edwards (New York: Putnam Publishing Group, 1999).
3. M. A. Runco, J. Nemiro, and H. Walberg, "Personal Explicit Theories of Creativity," *Journal of Creative Behavior* 31 (1997): 43–59.
4. Another good book to read on this subject is *Mindset: The New Psychology of Success* by Carol Dweck (New York; Random House, 2006).
5. Garr Reynolds, "Creativity, Nature, & Getting Off the Grid," available at: http://www.presentationzen.com/presentationzen/2008/06/postcard-from-oregon.html (accessed on October 3, 2013).

6. Thank-you to Christopher Walter, GIS Manager at the Cascade Land Conservancy, Washington State, for sharing the word-as-graphical separator trick.
7. James Cheshire, "Twitter NYC, a Multilingual Social City," http://ny.spatial.ly/ (accessed on October 31, 2013).
8. WNYC Data News Team, "Average Commute Times," http://io9.com/5988852/an-interactive-map-of-average-us-commute-times—how-does-yours-rank (accessed on October 31, 2013).
9. Damon Zucconi, "Fata Morgana," http://www.damonzucconi.com/show/fata-morgana (accessed on October 31, 2013).

Exercise

Pick three maps, digital or print, and analyze them in a short, three-page essay. Include small snapshots of the maps and discuss items such as unique features, colors, fonts, and layout in terms of their function, aesthetics, and efficacy. Isolate the color palette for each map (either using a software tool for getting exact colors or by approximating by eye) and include them—five to ten colors each—in the essay as small, in-line graphics. End with a discussion on which design elements are particularly worthy of adapting for your own future map efforts.

3

Layout Design

Layout design is marketing for your map!

All Together Now

The techniques in Chapter 6, "Features," are largely applicable to maps of all types, shapes, sizes, and output media. Imagine that you have just finished making the *best map ever* using those techniques and now you are ready to stick on some extras like a scale bar and north arrow, quickly send it off to the printer, or render some tiles and publish it to the web, then sit back and let your map audience admire your work.

Wait! Stop!

After all that hard work making your map, why would you want to risk publishing it without, say, a title or legend? And what if there are other things you ought to include there but didn't? Do you expect your audience to automatically know what this map is all about? I mean, of course the map itself is the most important part of your product and it is always a good idea to spend a lot of time on *that* aspect of your effort. But neglecting the *margin elements*—all nonmap elements—and their arrangement within and around the main map (or maps) can make your hard work appear confusing, unreadable, or just plain ugly. The layout deserves attention so that it can act as a showcase for the main map and also provide all the elements necessary to give the audience the appropriate context with which to understand it.

So what will you need to do to create a professional-looking layout? Following these points will help you get there:

- Look through a list of all the possible things that could go on the layout. You can find one in this chapter in the "Layout Checklist" section.
- Decide which of those things will go on the layout and get input from others.
- Research and gain inspiration from other maps and artwork, then choose a style.
- Decide which of the layout items will be emphasized and which will be understated.

- Decide on an initial configuration for those items and create an emphasis map or a wireframe mockup.
- Build the layout.
- Obtain feedback and repeat this process as needed.

Accordingly, the layout checklist and the element details and examples in the next section will guide you through the process for the first two steps. The last five bullets above consist of arranging the chosen elements into a stylistically cohesive layout by utilizing the arrangement principles found in the sections that follow "Element Details and Examples."

In order to give an overview of how a seasoned professional would create a layout, let's continue with the example of the all-too-hasty geoprofessional who made a great map, only to ultimately undermine that effort by not considering the layout. In this example, the geoprofessional is creating a static, printed map. Once it is printed, perhaps it looks so terrible that the geoprofessional decides to get out this book and follow the methods found here. So, to begin, she adds a title to the map after considering that her audience will need a quick introduction to the map that can be read in five seconds or less. She chooses a title that succinctly explains the subject of the map without using jargon. She then concludes that the scale bar and north arrow are fine, but that some text to document the data sources, author, and date ought to be included. Finally, after looking through the layout checklist, she is reminded of the need for an inset map that shows the study area location in a broader context for those who aren't already familiar with it.

Now that she has taken a first look at which items to include, she asks a colleague for some input. The colleague thinks there is a portion of the map that is too densely detailed for the map viewer to understand and recommends an inset map that shows a zoom-in on that area. With that added to the list, the geoprofessional begins the second phase of the layout method, the arrangement. She creates an emphasis map with a preliminary arrangement of the major items, decides that the map should be the first thing the audience sees, with the title coming in second, inset maps third, and the rest of the items last. After taking some time to arrange the elements in the geographic information system (GIS) software, the geoprofessional finally decides that the map is ready for printing. With some additional changes suggested by her colleagues, she revises it once again and then reprints the final product.

This geoprofessional managed to transform her layout from a novice-looking product to a professionally designed product by following the steps and details described in this chapter. Please note that she did not need to use all of the ideas presented in the following sections, and neither do you. Some of these ideas you may use so much that they become second-nature, and others you will use only when a particular layout needs them. You'll also note that the time that it took for her to complete her initial layout was a small fraction of the time that she spent creating her second layout. In fact, the creation of a professional layout can be extremely time-consuming, even

PLATING THE MAP

In cooking, the quality of the ingredients, the chef's skill in putting good flavors together, and the chef's knowledge of cooking technique and tools are all important and crucial. Even with all these elements in place, however, the chef must put effort into the presentation of the food on the plate to elicit the best reactions from the patrons. If the food is sloppily presented, the patrons often don't even taste the food! The best chefs, who by no coincidence also command the highest price, are those who put an extreme amount of effort into plating each and every order in the best way possible, no matter the cuisine.

In cartography, the quality of the data, the cartographer's skill in putting good data together, and the cartographer's knowledge of mapping technique and tools are all important and crucial. But the cartographer must also ensure that the presentation of that data is as understandable and aesthetically pleasing as possible in order for map readers to glean the required information from the map.

Thomas Keller, the famous chef and restaurateur behind the French Laundry and Per Se, said, "You eat with your eyes—you use your eyes first, so something that looks elegant and nice also looks appetizing."[1] Similarly, an aesthetically pleasing map is also going to be a map that communicates efficiently and effectively. Always challenge yourself to make the aesthetics of a map better for maximum map-reader attention.

for the most accomplished layout designers or user interface designers. Do not consider it unusual if your unique 8.5-inch by 11-inch masterpiece takes 40 hours to design, build, and print. A poster-sized layout could take anywhere from 40 to 200 hours depending on your level of fastidiousness, familiarity with your chosen software, number of people who provide input, and perceived audience size. (Notice I say "perceived audience size," instead of just, "audience size." The audience may indeed become much larger than originally anticipated, especially if the map is designed well. Some of the layout element descriptions, such as date and network path, will provide more detail on how you can set up the layout to provide adequate value to an unintended audience.)

As you read through the following sections, keep in mind that the main focus is on large-format layouts ranging in size from 8.5 inches by 11 inches to full posters as well as websites showcasing complex web maps. Printed layouts smaller than 8.5 inches by 11 inches tend to contain minimal to nonexistent margin elements. Digital, static layouts don't always need a lot of additional margin elements either, though it very much depends on the content and the map reader's familiarity with the subject. A choropleth map of illness rates in US states will require at the least a title, data source, authorship credit, and legend, whether it is displayed in print or online. In contrast, a very simple map of coffee shop locations in a small town may need no additional margin elements. The placement, styling, and best practices for margin elements on printed maps are well described here. Where possible, explanations pertaining to digital, interactive mapping elements are also described, though it is recommended that the reader also do additional studying on user interface design, in general, to supplement this, as many techniques that are needed in the digital realm are not limited to just cartography.

Layout Checklist

Primary Elements
• Title
• Subtitle
• Legend
• Maps
• North arrow
• Date
• Authorship
• Scale bars
• Page border

Secondary Elements
• Neat lines
• Graticules
• Network path
• Disclaimer
• Data sources
• Data citations
• Logos
• Graphs
• Photographs
• Graphics
• Map number, if series
• Tables
• Copyright
• Projection
• Inset maps
• Descriptive text

Element Details and Examples

Each line of the layout checklist contains what is referred to in this book as a *layout element* or simply an *element*. These elements include all the common ways of presenting information on a map and its surrounding layout. Any element that is not on the map itself is referred to in this book as a *margin element,* in reference to the fact that the element is located on the periphery of the layout. The map itself is also an element of the layout. Since the first step in designing a layout is to decide which elements will make the cut and appear on the final map, I suggest going over the included checklist to ensure that no element is overlooked. Any element that can lend additional informative support for the map ought to at least be considered during this step. Knowledge of these elements is important and should not be overlooked in printed or digital mapping efforts.

The questions to ask about each element include but are not limited to the following:

> Would the element provide information that is crucial for correctly understanding this map?
>
> > The answer to this question is not necessarily a no-brainer! Many times, we won't even realize that a certain graph would help explain the map better, or that a subtitle is needed, or that a photograph of each location will draw interest. First, of course, you will make the first cut. Carefully consider each element, try to put yourself in the audience's place, and also try to consider any "unintended" audiences such as future employees, nondepartmental colleagues, and so on. Asking for outside input is the next task. This is an important step that often makes the difference between a substandard map product and a well-made one. A colleague or boss can look it over and even your family can offer good advice. Soliciting a critique can be as easy as asking, "If I were to make a map of X, what elements on this list do you think would enhance your understanding of the subject?" To go a step beyond this and garner even more detailed advice, create a draft map before asking for input. Do not be afraid of critical feedback. Without it you can't improve nearly as quickly. The map is made for other people to look at, so why not involve the audience in the planning process? Present them with the layout checklist and have them identify any element that it is not currently present, but desirable.
>
> Would the element provide visual relief or create a cluttered feeling?
>
> > If you have checked every element on the checklist, then there are probably too many. An 8.5-inch by 11-inch layout will start to look cluttered by the time a fifth or sixth element is placed on it. A poster-sized layout could possibly support the inclusion of all the layout

elements if enough time is spent on their arrangement. (Time is generally what separates a good map from a poor map. Not experience!) Digital maps have more leeway, since the elements can be tucked away on secondary pages such as the "about" page. The main map page, though, should have the absolutely necessary informational items needed for the correct interpretation of the map, since most users won't click any of the supporting pages. Elements like logos and text are often present on maps, but this does not mean you have to use them yourself. If they generate clutter and have no functional value, then you can safely and assuredly get rid of them (unless a situation arises which requires a logo or trademark).

Is it necessary?

Carefully consider the necessity of each element. For example, you may think everyone knows where north is because your entire work group is more than familiar with the area in question. If your map's audience is your work group, then you are more than justified in omitting a north arrow. However, if there is an inkling of possibility that someone from "the outside" will view the map, perhaps it wouldn't hurt to allow an unobtrusive, simple, north arrow as standard practice. I caution, though, that you have to decide when standard practice verges on the ridiculous. For example, a disclaimer on every map that your department creates is a nice idea *most of the time*. But when you are passing out directions to the ice cream store where you are holding a staff meeting and are required to include a disclaimer that takes up half the map, it could get silly. Another silly situation is requiring a user to read and agree to a disclaimer—especially disclaimers of more than one paragraph—before they are allowed to view the map online. This is a sure way for one of two things to happen: the user will not read the disclaimer but click "agree" anyway, or the user will lose interest and never look at the map. In other cases, standard practices aren't adhered to enough. Often we will see maps without titles that should have titles. One should not have to guess at the subject of the map or, perhaps worse, have to look at the legend in order to discover the main message.

Once the decisions about which elements to include and why have been made, the next step is to decide where they will go with consideration toward functional and visual appeal. Many of the nonmap layout elements can be placed either on the map itself or in the margins depending on your design schema, map characteristics, and other considerations. While those are by far the most important aspects to layout design (the *what* and the *where*), you can go further in your pursuit of layout excellence by carefully considering the design of each of those elements. You'll need to think of each in terms of color, font, word choice, and overall style. The details for each element that follow will therefore not only help your decision as to *what* to put on the layout but also help inform their placement and their style.

Title

Though a title is short, give it lengthy thought. Remember, the title's purpose is to succinctly pronounce the intent of the map. Making sure that this is the case takes a bit more time than just throwing on the first few words that come to mind.

Best practices: Take a look at Figure 3.1 and decide which title you think is better.

You'll often see map titles that identify the authoring agency as well as the geographic location of the map. It is not usually necessary and often unintentionally arrogant to include the authoring agency or company in first position of the title. A better approach is to include the information in the subtitle if it is deemed sufficiently important or, more advisable, to include it in the authorship element (see the "Author" section). Similarly, think twice about putting the name of the geographic location in the title. The geographic location ought to be immediately readable from the primary map element and, if needed, in its overview map element. This renders the geographic-location-in-title practice redundant and, well, boring. The exception is when the map does not display an analysis of any sort, thus making the geographic location one of the main focuses of the map such as in a road map titled, "Larimer County Roads."

The title is either the primary or secondary layout element (if secondary, it is only second to the map element). It is written in large enough type to grab attention and be readable from a generous distance. On web maps, in particular, it may be the only thing the map reader will pay attention to other than the map itself. It is short enough to read quickly, it is interesting, pertinent, and accurate. It ought to summarize in ten words or less the primary finding of the analysis, if it is a thematic map. Avoid any obvious and redundant terms such as "map of ..." or "analysis of ..." Also avoid using jargon such as "framework" or "model," as these result in useless mind-clutter for the reader. All words in a title should be spelled out, so avoid acronyms at all cost. Attention-grabbing tactics such as questions ("How Much Purple Toad Habitat Do We Have?"), sensational assertions ("Purple Toads Living Large!"), and

Schmoe County Land Consortium's Analysis of Purple Toad Habitat in Northwest Hooktown

OR

Purple Toad Habitat Greater Than Previously Thought

FIGURE 3.1
The first title is long and laborious and is suitable for dry analytical presentations. The second title gets straight to the point and is suitable for all audiences.

actions ("Improving Purple Toad Habitat") can serve the map well at conferences, in office hallways, cubicles, news sites, blogs, and many other locations.

Placement: Typically, a title is located at the top or bottom of a layout and is either centered or flush left. Less common, but also acceptable, are just above the legend or titles that are oriented vertically, headed upward along the left-hand side of the page. If placed inside an architect's margin element box (see the "Margins" section), the title is generally shown at the top of the box, centered, or in the case of a vertical right-hand box, flush left. The diagrams in Figure 3.2 show some title placement options. The title is depicted in these diagrams as a gray bar.

Style: Using all capitals is acceptable in a title; however, consider a small-cap style for a slightly more readable look, especially if using a bold font in conjunction with the uppercase style. A drop-cap for the first letter may also be used. Some argue that words in all upper case are never okay because people cannot easily decipher the letter codes when the letters are all the same height. As the argument goes, lowercase letters are read easily because of the different heights of various letters, resulting in specific shapes for each word. These shapes allow a reader to scan a word rather than look at it letter by letter. This makes words that are in all capitals, and therefore all the same shape, harder to read, resulting in more time taken to read the same words. However, a title on a layout usually needs a lot of emphasis in order to attain its proper placement as number one or two in the hierarchy of layout elements. Often, the all-capitals style is the only way to achieve the amount of emphasis needed. Additionally, a title containing ten words or less ought to be short enough not to cause any undue reading strain on your audience, regardless of the uppercase style.

FIGURE 3.2
While titles could be placed anywhere on the layout page, these placements offer visibility while still allowing the map element to remain central to the overall layout.

Estimating Impervious Cover Under Full Buildout

: Riverine Habitat Inventories

Best Sites For Ground Mounted Solar Systems

Fast Food Chains Per Capita

D I S A P P E A R I N G B I R D S

FIGURE 3.3
There are many different ways to bring a title to life. Shown here are several options ranging from simple to complex. The last title uses expanded character spacing, which requires a serif font for better readability (see Chapter 4, "Fonts"). The other titles are in a sans-serif Arial font.

There are some other techniques you can use instead of, or in addition to, the all-capitals style, including small-caps; drop shadows to lend additional visual weight; special leading characters such as a pipes (|), leading dots (...), or colons (:); drop-caps and extra-large first letters; and underlining. In terms of color choice, dark grays are being used much more now and are a great alternative to black, which tends to be more obtrusive and harsh. Therefore, a bold dark gray can make a modern statement as long as the title retains its proper level of emphasis. If it still doesn't seem bold enough, then switch to black. Also, avoid using bold red as it tends to contribute to a dizzying Willy Wonka effect. The point is to provide sufficient color contrast. A dark background, obviously, will require a light text color: either bright white or a gray-white, typically. Some style options are shown in Figure 3.3.

Subtitle

The subtitle is comprised of any spillover text that is slightly less important than the title but still somewhat necessary to fully understand the map. This is a better place to put the geographic location of the map and the sponsoring or authoring information than in the title. But again, if possible, leave that type of information for metadata text blocks in less conspicuous locations of the layout rather than the subtitle. Ideally, the subtitle provides further detail that the viewer needs to know prior to being able to understand the map.

If your title does not state a key finding or impressive fact, then this information could also be placed in a subtitle. For example, the title, "Channel Conditions Database Developed for In-stream Data" could have a subtitle such as, "Data from multiple sources and protocols integrated seamlessly into a single comprehensive spatial database." In this case the *what* is described in the title and the *why* is described in the subtitle. Another example would be a title and subtitle pair such a, "Species Distribution Change Over Time" subtitled with, "Five Species Tracked—Four Show Significant Change." You notice that unnecessary words are discarded for the sake of conciseness and that the subtitle here is relating the key finding of the study. A nonanalysis-type map might use a subtitle to relate an important qualifier for the map such as this title and subtitle pair: "Animal Shelter and Rescue Organizations" subtitled with, "Non-Homecare Facilities Only."

Best practices: The subtitle is displayed in the same font as the title but carries slightly less emphasis than the title. This is accomplished by not using a bold font on the subtitle, employing a smaller font size than the main title, and perhaps italicizing or indenting the text to further deemphasize and separate it from the main title.

Placement: The subtitle is placed directly below or to the right of the title.

Legend

The legend is a standard element on most layouts. It provides the color and symbol key look-up details for the map element. It is comprised of items (icons, points, lines, polygons) and their associated labels (descriptions of the items). If the layout is being created for a narrow audience, such as your workgroup, and the map includes commonly understood feature types with standard symbology such as county boundaries, water bodies, elevation, and so on, then these may be granted an exception from inclusion in a legend. Wider audience maps may also exclude certain given feature types drawn with standard symbology such as blue water bodies or green land expanses; however, these are left to the discretion of the map author. Only exclude map layers from a legend—or the legend itself from the layout—purposefully. Don't just forget some! Err on the side of the more legend items the better since a map is not worth much if its features cannot be understood. What is obvious to the map author can easily be unknown to the map viewer.

In some situations the entire legend could conceivably be omitted. For example, a map layout on a slide doesn't need them due to the fact that the presenter will be talking about the slides and can point out the necessary features. Legends are also often unreadable on slides presented to a large audience. Be cautioned, though, that many slides are reproduced online or in other media, and these will need legends, especially when they are not accompanied by the presenter's notes. Another situation in which you could omit a legend is when you are creating report maps that comprise a series of similar features. In this case, one legend at the beginning of the series can apply to all the maps and thus allow more space for the map elements on the subsequent pages. Simple, small, report maps may also have so few features that they only need to be explained in the text of the report or in the map captions. Exceptions also include basic basemaps with standard symbols or interactive maps that are well annotated or utilize click and hover techniques for point and area-based data.

Best practices: Items look best when placed to the left of their corresponding label. Complicated legends with many items necessitate grouping levels.

The two forms of grouping levels most commonly seen are the categorical group (e.g., all land-use colors placed separately from all stream-level colors) and the shape-type group (e.g., all polygon features placed separately from all line features). Use headings to describe each group in categorically separated legends, if your organization schema is not immediately obvious. When categorical separations are not needed, shape-type groupings are often displayed in the following order: points, lines, polygons. Other times they are displayed in the order of relevance to the overall map purpose.

Default legends produced in the GIS are a great way to begin production of the legend but ought to be further manipulated prior to map publication. First of all, make sure there isn't too much space between the items and their labels. Also, if the legend is describing a gradient of values (choropleth color scheme), then those items will look more professional if they are touching each other rather than separated by white space. Don't forget that color gradients can be depicted vertically (Figure 3.4) or horizontally (Figure 3.5).

A label can still be associated with each item, although in many cases we can do without the intermediates and label only the maximum and minimum values. Creative color gradient legends could also employ other effects such as dials (e.g., speedometer-type dials) or graphs (e.g., elevations and colors).

FIGURE 3.4
Color gradients depicted vertically.

FIGURE 3.5
Color gradients depicted horizontally.

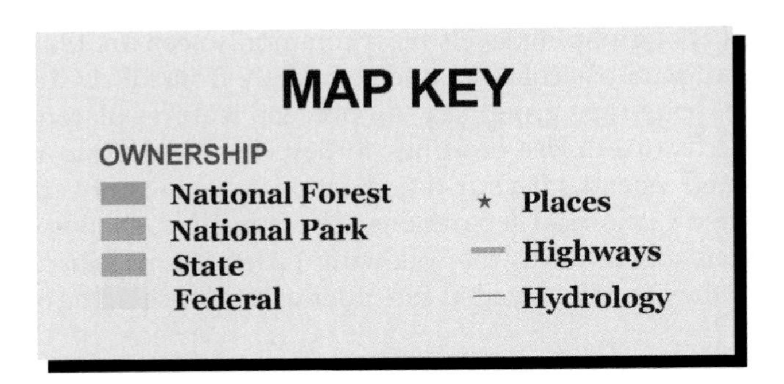

FIGURE 3.6
A horizontally oriented legend is shown here.

An often overlooked feature of the default legend is the default legend font. If a legend is created with the software automatically, it will use a default font, which needs to be changed to match the other fonts on the map so that a cohesive style is maintained.

As with gradient-type items, the groups of items in a legend do not need to be arranged vertically if it would fit better on the page to distribute them horizontally. The legend in Figure 3.6 illustrates a horizontally oriented legend.

A legend title such as "legend," "symbols," "key," or "map key" is not necessary.

Placement: Legends need to be within or nearby the map element with which they are associated. Legends can be placed outside of the map element, in the margin, in either of the following two cases: if there is only one map element on the layout or if all map elements on the layout share a common legend. If there are multiple map elements with different features on each, it is not always clear which legend goes with which map unless the legend is superimposed on its map element. When placed inside the map element, it should not obscure the underlying data and is often therefore placed in areas that are not important to the map's purpose such as on the ocean for a terrestrially focused map or on the land for a marine-centered map. In such instances, a background box is often needed to provide uniform background color on which to set the text. Online, interactive maps can perform well when the legend is located at the top, center, of the page. For example, a global temperature map could show the temperature gradient and associate labels in a horizontal bar at the top of the page in place of a title. These types of legends for interactive maps may themselves be interactive, such as time sliders in which the user changes the time period for the map's data via the legend, which then updates the map. Including this kind of interactivity is a great way to get the user to stay on the page longer and explore the data to a greater extent.

Style: When the legend is placed outside of the map element on a static map it can be encased with a shaded box or outlined box. However, it may look better if the outline and shading are done away with so that the legend can be incorporated with the other margin elements. This creates fewer seams on the layout. If a separation is still desired, a compromise is to use a shortened line above and below the legend (see Figure 3.7).

Under certain circumstances, you might use the same background color for your legend that you use in the map element to ensure that the colors will look the same in the legend as they do on the map. For example, let's say you have a map with some buoy locations shown in yellow on top of blue water. For the map's legend, though, you've chosen a light yellow, almost tan, background color. If you try to superimpose the buoy color onto this light-yellow legend background, you'll see that the buoy color doesn't look nearly as vibrant as it does on the blue water background. It may even look like a different color entirely (see Figure 3.8). If you find this sort of thing happening on one of your maps, the best thing to do is to change the background color of the legend to something that more closely matches the main background color of the map. This effect, where colors appear differently depending on their background

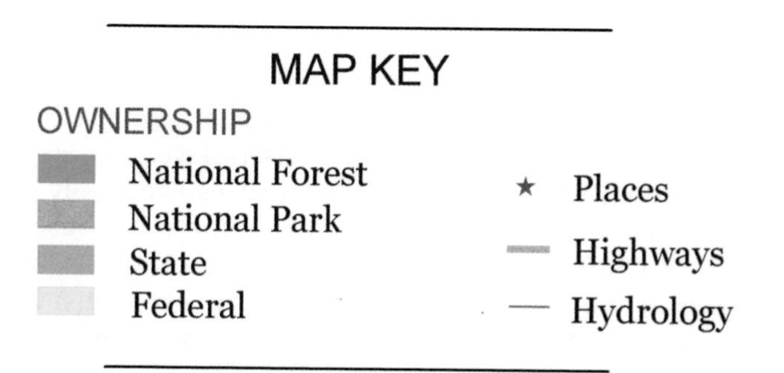

FIGURE 3.7
Rather than creating a full box around the map key, a short line above and below the legend items could suffice, keeping things neat but uncluttered.

FIGURE 3.8
This is an example of chromatic contrast. The two inside boxes are actually the same yellow color, but it looks different depending on the background. This becomes important when you are trying to match legend colors to map colors.

color, is called *chromatic contrast*. The same thing can happen with grayscale colors, but it's then called *achromatic contrast* or *lightness contrast*.

As far as the legend labels go, try changing the style of the text around a bit to add some interest and variety. For example, if you are using a graduated symbol to represent numerical ranges, you could go with a common format such as: "0–10%" or you could try a slight change such as: "0 to 10%."

Maps

The map element is a large graphic that shows off data in coordinate space. There are two questions that you need to answer as soon as you start thinking about a layout. First, how many maps do you need on one layout? Second, should you overlap multiple data layers (i.e., datasets) onto one large map element or separate the data layers onto multiple smaller map elements?

Multiple maps are generally used when displaying the same data over separate time periods (in such cases, a common scale and legend are important) or when displaying differing geographic extents that all have related data. Additionally, multiple maps may be required when ancillary data that contribute to the overall map purpose need to be displayed but are not central to the main purpose. Multiple maps on one layout can be highly informative as they offer on-the-fly analysis by the viewer of the map, thus making the map interactive and more likely to endure in the viewer's memory. Whether your layout will contain one map or many, the map element(s) will be your primary means of conveying information to the viewer, and therefore will likely be the layout element that takes, by far, most of your time to create and polish. Indeed, Chapter 6 of this book is devoted solely to the techniques used in developing and refining the map element. It can be designed in conjunction with the rest of the layout or it can be designed prior to the layout design process.

Best practices: Because Chapter 6, "Features," describes in detail the best practices and other design considerations needed to create a map element, this section focuses solely on general design considerations for map frames and geographic extent. A printed map or maps can be framed with a border (either simple or fancy) or it can be left to "float" in the layout space without a border. A studied look at many of the most impressive, elegant, and recently designed maps will reveal that for the most part their designers have included no map frame. In contrast, many of the GIS maps made in the 1990s and 2000s were made with these frames. Usually, there is such a stark contrast between the map and the surrounding layout that there is no need to provide the visual separation that a border provides. For example, a map that is completely covered in various colors put onto a layout with a white background already has a built-in visual separator in the white space between the map and the other elements. Similarly, with online maps, the white space surrounding the map element is sufficient to separate it from menu bars and interactive data layer pickers. Indeed, with online maps, it's often the smaller margin elements that, by means of a contrasting background, have the separator boxes and not the map itself.

In static mapping, when we are faced with presenting the results of an analysis, we often wonder if the analysis extent should be the same as the visible map extent or if we ought to include the surrounding geography in order to provide context, even though it may detract from the central focus. For example, let's say you've analyzed which households will be impacted by a proposed tax increase within a town's boundaries. Do you display just the town, given its irregular border, or do you show the houses that lay outside the town boundaries as well, out to such an extent that the map becomes a square or rectangle shape? There are a few techniques used to solve this problem. You can clip out all the outlying areas, leaving the map to "float" within the rectangular or square area that the page requires. You can include the outlying data as-is. You can include the outlying data in a generalized form. Or, you can provide context while still maintaining the focus on the analysis extent by changing the outlying data to a faded or semitransparent look. This last technique gives the analysis extent a popped-out effect.

Placement: In most cases, the map element or elements will consume the majority of the layout space. When there is a hierarchy of maps on the same layout, there should be no doubt as to which is the main map. It is either set apart in size or position, or most effectively, both. When just one map is desired, it is placed in a central but slightly off-center location on a static layout. An interactive, digital map layout looks fine when the map is central. Multiple, static, time series maps that are all the same size and shape function similarly to a single large map in terms of layout placement. Do you have to conform to these conventions? Not necessarily, but make sure you have good reason for your decision. For example, there are layouts that contain a centered map with logos or pictures surrounding it on all sides, such as city tourist maps that are distributed for free. These maps often showcase advertisements around the map on all sides. Because the advertising is very important on that type of layout, the placement can be acceptable, but on most analytical and informational layouts, surrounding the map with an equal amount of margin information on all sides results in an undesirable amount of clutter and lends itself to confusion as to the main focus of the layout.

North Arrow

The north arrow, whether it is fancy or unadorned, has the sole purpose of illustrating the orientation of the map to the viewer. Yes, it is true that most GIS maps are already oriented with north at the top of the layout. However, there are certainly instances where this is not the case and therefore it is standard practice to include the north arrow. Its inclusion on static maps is almost always warranted. Nautical charts and other orienteering maps should show both true north and magnetic north. In fact, these types of maps usually illustrate the directions on a compass rose, rather than a north arrow, which shows at least four, and sometimes more, cardinal directions. Do not include a north arrow on any map that uses a spherical projection, such as Mercator-based web maps. North is not a constant in those projections and must instead be indicated with graticules.

Best practices: Keep the north arrow small, simple, and unobtrusive for most modern layouts. Historical visualizations or other unique situations may warrant a fancier, bolder look.

Placement: North arrows are best left in a less conspicuous area of the layout. Sometimes you might use one to conveniently balance out some other element. Grouping the north arrow with other ancillary map information such as the scale bar and legend is another common practice. In a very small layout meant for a slide or in an in-line report graphic, the north arrow can be placed directly on the map element, in a corner, perhaps with a box of contrasting color behind. This is a technique used sometimes in 8.5-inch by 11-inch layouts and more rarely on larger layouts. The larger layouts tend to display north arrows outside of the map elements and grouped with the scale bars and other supporting information.

Style: For some reason, we geoprofessionals (and, indeed, cartographers as a whole) have not developed one standard north arrow that everyone uses. Instead, there exist hundreds of the things out there to choose from, in all manner of styles. And even with that, geoprofessionals still sometimes get the idea that each company or department ought to develop its own. It is almost like a logo in these cases. And to this I say, well, why not? It is fairly easy to design one yourself and it will further help your cause to put a unique stamp on your map.

As mentioned earlier, the trend in mapping design has been going toward simplistic north arrows for some time. The context and style of your overall layout will help determine whether you should stick with this trend or branch out to a fancier style (perhaps you are displaying historical data, an archeological dig, or some other type of map that may lend itself to a less modern-era north arrow).

Date

The date referred to here is the date on which the layout was printed. (Dates of your data sources are addressed in the Data Citations section). It is important to include the date on most layouts that are intended to be stand-alone prints. Layouts destined for reports do not necessarily need a date because the report ought to already contain date information, although it is still recommended to include one should there be any chance that the map will be copied separately from the report. Layouts destined for slide shows also do not need a date for the same reason, although one could be included if the slide show will be posted in a digital realm or if your map slide will be separated from the title slide (or wherever the date is). Interactive maps don't need dates as they should be updated when needed and are thus assumed to be current.

Stand-alone layouts, such as 8.5-inch by 11-inch sheets and larger, should include date information as a form of embedded version control (e.g., "Oh, I see now, that was the old map I printed out last week; here's the one from this week."). The date also gives the audience an idea of the map's vintage for maps that endure. Whether or not the layout designer considers the map an enduring one is usually inconsequential. Maps often contain items of value that are referred to long after the mapmaker originally thought possible. For example, the map you created only for this week's board meeting might take on a life of its own by getting passed around (digitally or on paper) after the meeting, and before you know it you have created a lasting legacy map that people have photocopied and put in their files. Let's hope you thought to put a date on it so that those who will pull it out of their files and "dust" it off will instantly understand whether it is of a useful vintage or not. In summary, inclusion of the date on a map or layout is often a preemptory practice used to mitigate unforeseen events.

Best practices: The date should be preceded by text such as, "printed on:" to dispel any confusion over whether the date refers to the data or the day of printing.

Placement: The date is considered metadata and therefore should be placed with the other margin elements in an inconspicuous location.

Authorship

Authorship refers to the person and agency who designed the map, created the analysis, and put together the layout. It can include more than one author if these tasks were completed by several people, but it always includes at least one name, not just the agency. Authorship information is not to be confused with logos, which don't necessarily state who created the map. For example, does that logo near the title signify that that was the company that contributed the underlying data, performed the analysis, designed the layout, or sponsored the work? Who knows? Authorship is a plain text way of getting that information across to the viewer of the map. The agency that commissioned the map or is in some other way involved in the map's reason for being can be signified more prominently than the authorship information, as long as the author's name is included somewhere on the layout. Indeed, many times the sponsoring agency's name ends up in the title of the layout, although I do not consider this a best practice (see the "Title" section).

Authorship on web maps is a must. For some reason, a great deal of interactive web maps don't have authorship information. This makes citing the maps in any derivative works or referencing them in other works difficult to impossible. Another poor practice is citing only one's cryptic username rather than the real name. In some cases the authorship is included but it's hidden in less-than-obvious "about" page links, which only serves to encourage the citing of the map without anyone knowing who actually created it. A map of serious scholarly consequence, whether it's on paper or a digital device, needs real people standing behind it.

BEST PRACTICES

According to Edward Tufte, in his book *Beautiful Evidence*, the author's name is important to include for many reasons, such as signifying that someone is taking responsibility for what is contained in the printed work, providing the contact name for follow-up questions, and signaling reputation (or lack thereof). Furthermore, he states, "Authorship credit is too often absent from corporate and government reports; we should remember that *people* do things, not agencies, bureaus, departments, divisions. People may do better work when they receive public acknowledgement and take public responsibility for their work. The good [Charles Joseph] Minard put his name on nearly all his work and personally signed with pen and ink ... some of [his] ... figurative maps."[2]

Placement: Conference posters can include the authorship information directly after the title and subtitle, or in the descriptive text portion of the poster, or in a corner by itself. If placed under the title(s), we will often see the author(s) names followed by numbers in superscript that are referenced elsewhere on the poster for the author's affiliation and contact information. On smaller maps and even some poster-sized (nonconference) maps, my preference is to put the authorship information in a corner, in dark gray italicized text. In this way it is clearly metadata: it is there if the viewer needs it but does not intrude on the layout. Digital maps can include the authorship in a similar fashion or on a separate page as long as the navigation to that page is clearly apparent.

Style: The authorship information ought to include as much information as possible including organization, address, phone numbers, email addresses, and web addresses. In print, this information can be organized with comma or pipe separators (|) to ensure a neater look than hard returns between each item. Incidentally, a practice that is sometimes used in architectural renderings is to have certain supervisors sign off on the completed design. This could be a useful addition to a GIS map being used in a workgroup situation. For example, if the map needs to be examined by several supervisors, you might type out their names next to several blank lines to serve as signature lines.

Scale Bars

A scale bar is a graphic that shows the map viewer how to translate between map units and real-world units. While historic maps often had verbal scales such as, "200 feet to the inch," or "1 inch = 200 feet," the modern convention is to use a graphic scale. When a paper map is scanned, shrunk, and put on a slide, or some other similar enlargement or reduction process takes place, a graphic scale will remain accurate. Scale bars are suitable for most large-scale maps but not suitable for maps where distance and area are not constant.

Best practices: Modern scale bars are simple and unadorned. In many cases the viewer needs to see only two divisions—one at the beginning and one at the end—and a number indicating the distance between those divisions in real-world units. This is due to the fact that the typical GIS map is not used to measure distances. The scale bar, therefore, serves only to illustrate the general scale of the geographic extent. Subdivisions aren't needed. The point is, if you are creating a simple map, stick with a simple scale bar. For example, let's say you are making a 3-inch by 3-inch graphic for a newspaper story that shows general school-district boundaries. You do not need to create a huge scale bar showing 4 major and 4 minor divisions. A single line with labeled end points will suffice. Even with simple scale bars, however, there is utility in providing multiunit scales together on the same map so that the viewer need not perform mental unit conversions should a different unit be desired. Of course, maps that are used for the primary purpose of pinpointing an exact location or distance, such as hiking maps or road maps, do require a more detailed scale bar that shows subdivisions and their associated measurements.

Style: Scale bars are almost always black, or in some instances, dark gray, in color. The font used should conform to the other fonts used on the layout. Several layout styles are possible depending on the space allowed within your layout. The more compact form of the scale bar is used more often on modern maps, whereas the more linear form is a relic from previous decades (see Figure 3.9).

Either format is acceptable, though one should tend toward the compact form since it is more balanced and modern. As mentioned earlier, multiple unit scale bars are sometimes required or beneficial. In this case, we can simply show two separate scale bars with the different units, but place them close enough to one another to maintain an orderly looking layout (see Figure 3.10).

You might also want to note the map scale as a representative fraction (e.g., 1:24000) in the same visual space as the scale bar as in Figure 3.11.

An uncommon way of depicting scale, but still worthy of note if the data is especially suited for it, is to show the scale in area form. Many population density maps, for example, are shown in square mile or square kilometer units and would be ideal for this kind of scale graphic. Additionally, a map with buffers around points at certain distances could also use a scale graphic. See Figure 3.12 for examples. If these types of scales are used, they ought to be shown in conjunction with a conventional scale bar as well.

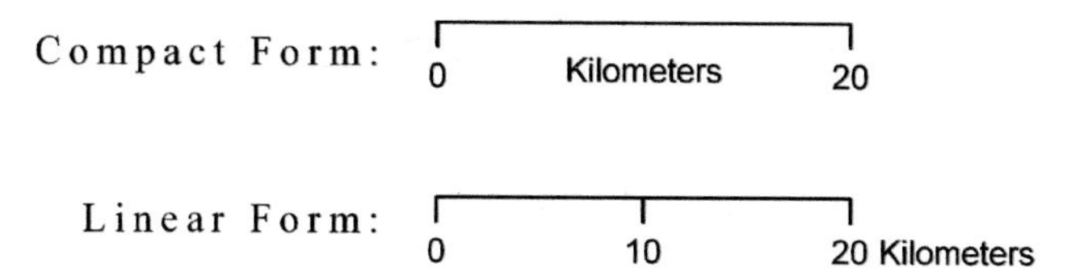

FIGURE 3.9
Two formats for scale bars are shown here. The compact form is more modern, the linear form is classic.

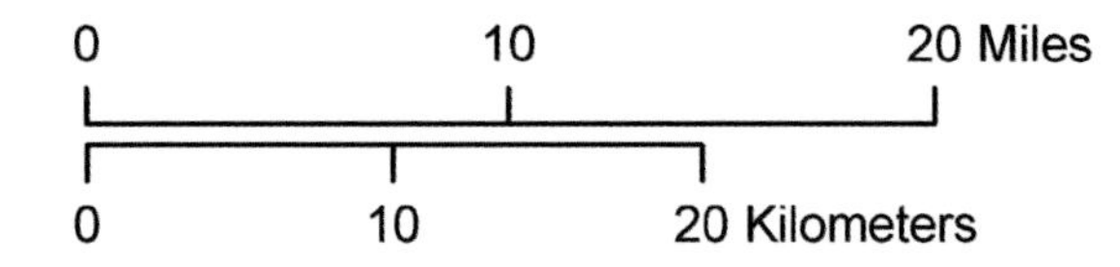

FIGURE 3.10
When two scales are required, place them close together like this.

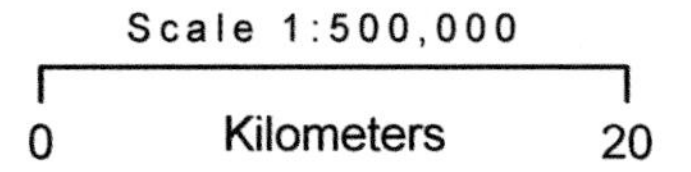

FIGURE 3.11
A scale bar coupled with a representative fraction.

FIGURE 3.12
Other scale types can also be depicted on the layout including square-area and buffer distance.

Page Border

A page border is used to group all the layout elements of a print map together by means of a single graphic line surrounding all of the elements. When the layout consists of a single map element with all other elements floating on top of the map, the page border is sometimes, but not always, redundant with the map border. When the layout contains elements outside of the map element(s), a page border is usually used to contain those items.

Placement: The page border is placed around all the layout elements, including a narrow white-space buffer that acts as a frame. The frame does not have to be white; it can be any color that unifies the layout. The frame ought to be the same width or proportional to the other separating spaces on the layout.

Style: Double lines, single lines, varying thicknesses, shadow boxes, rounded corners, and so on are all used and acceptable. The simplest and best is the single line page border so as not to detract from the surrounding elements. Just make sure that poster-sized maps at C size or larger that have page borders that are of at least 3-point thickness, and up to 5-point thickness for E size sheets.

Neat Lines

A neat line is simply a graphic line placed on the layout. These lines can be used singly, in tandem, or in groups, and can consist of simple lines or boxes. A large neat line around the entire layout is referred to as a *page border* and is discussed, fittingly, in the "Page Border" section preceding this one.

FIGURE 3.13
The architectural margin boxes shown here organize the margin elements into the same three sections but separate them differently. The top one achieves the same goal of visual separation but with much less visual clutter than the bottom one.

The purpose of a neat line is to explicitly separate elements to provide an organized look. Sometimes the same visual relief and separation can be achieved without neat lines by simply utilizing the empty space between elements for this purpose. However, through experimentation with your elements as you place them on the page, you may feel that they need more separation than the empty space provides. For example, a poster-sized layout with three major sections in the architectural margin element box (which is itself a form of a neat line) may benefit from the use of a short vertical bar between sections. The reason a short vertical bar would be used instead of, say, a box around each section, is to reduce visual clutter but still provide the required separation (see Figure 3.13).

Placement: One neat line placement illustration is in the Legend section of this chapter (see Figure 3.7), which shows a neat line placed above and below a legend in order to separate it from other elements on the layout. Experimenting with different places to put neat lines on your own layout, as well as getting ideas from existing layouts, will instantly increase the quality of your finished product. In fact, even though neat lines are on the lowest end of the information spectrum for map elements, they are on the high end of the scale of the design spectrum for layouts. It is these little touches that can make a map look like it was created by a professional with years of experience. Although be warned that it can also take years (slight exaggeration) to place, re-place, and tweak these so that they look just right.

Style: Neat lines are almost always black or dark gray when placed on lighter-colored backgrounds. The thickness of each line needs to be commensurate with the importance of the information it is enclosing or separating, as well as the total layout size. Neat lines are drawn as boxes or simple lines. If using lines, and they are intended to meet up with other lines, ensure that they join together neatly by using a snapping feature and by zooming in to the largest extent to double-check the results. Sometimes what looks like a snapped line winds up printing out as an off-shoot, like the top example in Figure 3.14 instead of properly snapped like the bottom example in Figure 3.14. Conversely, ensure that any neat line that is *not* purposefully intended to meet up with another line is far enough away from all other lines so that it does not look as if it was supposed to be connected.

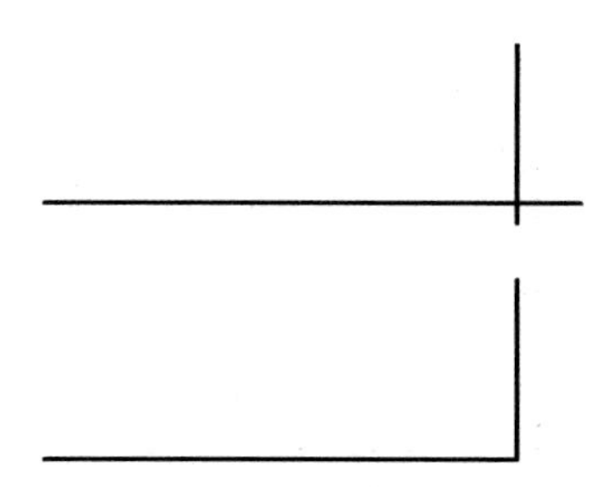

FIGURE 3.14
Make sure any lines that you use to organize layout elements snap together appropriately, as in the bottom part of this example.

Christopher Walter, the GIS Manager at the Cascade Land Conservancy, came up with a unique twist on the typical neat line. Instead of using a line to separate blocks of text in his map's margin box, he used *words.* The words were written in a horizontal fashion, reading from bottom to top. For example, between the page border and the legend he wrote "Map Legend" and between the legend and the data sources he wrote "Data Sources." In this way, he titled the separate parts of the margin and separated them all at the same time.

Another symptom of neat lines not ending up quite so "neat" as intended is when they show up as jagged lines on the printed page. The cause of the zigzag is the software's habit of starting and ending a line at exactly the start and end points that were clicked when the user created the line; if those points are not on the same axis, a jagged line is created. Use guidelines and rulers when possible along with snapping functionality to avoid this often undetected issue.

Graticules

Graticules are latitude and longitude lines that run along the surface of the map element and enable the viewer to visualize how the flat map surface relates to the real-world three-dimensional (3D) surface in the map's projection. Although this is an arguable convention, graticules are almost never present on GIS map elements that are not for navigational purposes. Even so, you may consider including them on your nonnavigational maps as an additional location- and scale-related metadata element. Maps with nonconstant direction (i.e., north is not always directly "up," such as those in the Winkel triple projection, for example) are well-served to have at least a few graticule lines to indicate direction.

Best practices: Layouts showcasing analytical-results maps do not need to include graticules, and indeed, if they are included, they may provide too much clutter for the map audience to decipher the analytical results easily.

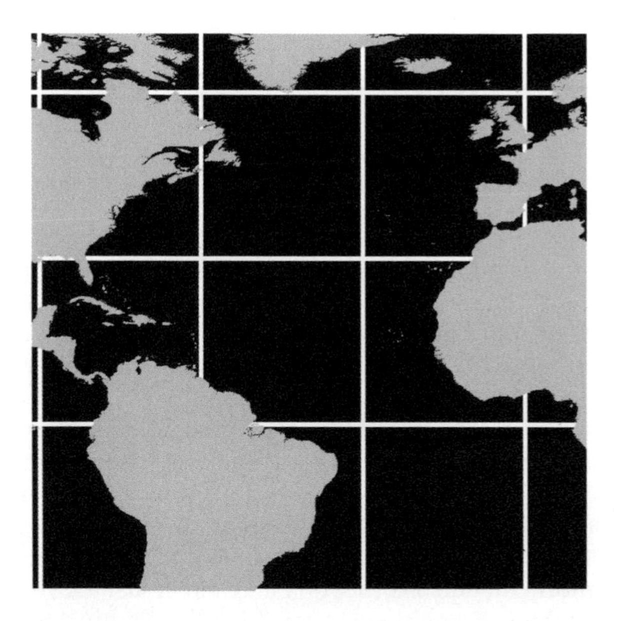

FIGURE 3.15
A common convention is to place the graticules beneath land masses when designing very small-scale maps.

If they are still wanted, make them thin, dashed, and/or few and far between. Some small-scale maps might only show the equator, the Tropic of Cancer, and the Tropic of Capricorn, for example.

Style: Navigational maps will require easy-to-read, prominent labels at the top and bottom of the graticule lines. If you want to include graticules on a nonnavigational map, you will likely want to make them unobtrusive on the map surface by using gray or dashed lines, fewer divisions, and small labels (that are visible if the viewer looks but do not intrude on the other important aspects of the map). An additional technique to minimize their interference is to run the graticules behind certain portions of the map, usually the land portions of a worldwide map (see Figure 3.15). With this technique, you can either create graticules that intersect the continents or create graticules that stop just prior to intersecting the continents. An "invisible" buffer around the continents can accomplish this latter technique. For this, you buffer the continents, layer the buffer on top of the graticules, and assign the same color to the buffer as you are using for the ocean water.

Network Path

The network path (e.g., \ProjectResults\AnalysisB\ConferencePoster) of a printed map can be included on the layout as part of the metadata textbox or as its own entity tucked away somewhere even less conspicuous. Its purpose is to alert the viewer to the location of the layout and data files should

\\ServerName\Share\ProjectResults\ConferencePoster

FIGURE 3.16
A text item directly beneath the page border or hidden in a corner like this signifies to your audience that the text is for internal reference.

the map need to be updated or the data be used for a future project. This is one of those items that is hardly ever seen on a layout, although it should be included on most internal layouts and ought to at least be considered for external layouts. Network path can also refer to the path to the public repository or private cloud-based data storage location.

Best practices: Depending on the complexity and number of projects in your GIS workgroup, you may want to consider making the inclusion of network paths a standard practice for the printed maps produced by your workgroup.

Style: On paper maps, the technique used for nonessential but informative data, such as the network path (and sometimes authorship, date, and so on), is to put the information in a corner of a map either just above or just below the page border. Thus it would look something like Figure 3.16.

Disclaimer

The GIS map disclaimer is used by most public agencies and many private companies to protect themselves from lawsuits arising from the map being used for unintended purposes and to inform the reader as to the potential limitations of the map product. For example, disclaimers often contain text such as: "the agency provides this data *as is,* it is the user's responsibility to determine proper uses for the data; the agency is not responsible for any adverse outcomes associated with such use; and the features are not at a survey scale and are not complete." If you are tasked with creating a disclaimer, then it is wise to look at the myriad examples that exist on the web as well as to get input from your legal counsel. Those who work in agencies where the disclaimer is already written need only figure out how to best place it on the map and style it appropriately.

Best practices: Whether or not to use the word *Disclaimer,* or some other title, at the beginning of the disclaimer text block should be considered. Does it really need to be there? Could a term like *Note* or nothing at all work instead? On the flip side, you may want to add a term like *Standard Disclaimer* or Disclaimer of Liability in order to alert the viewer that this is not a special block of text that is only used on this one layout. In any case, you may be constrained by what your legal department requires.

Style: Digital, interactive maps sometimes present the user with a disclaimer that must be clicked away before the user gets to the map. This practice is discouraged because it drives users away before they get to the map, and those who do click through have, for the most part, not read the disclaimer. It is a much better idea to include the disclaimer as a text element over the web map in a corner or to include it in an "About" page. On paper maps, the common practice of printing the disclaimer in small, but legible text on the layout with the other metadata elements is pretty much the only way to go. Unfortunately, the disclaimer can be quite long; resulting in a huge text box relative to the other elements, and it is left to the layout designer to figure out how to print such a large block of text in an unobtrusive fashion. Using a gray font color or less than single-spacing between lines may help. Also, pushing it to the edge of a layout as opposed to placing it between two other margin elements may help minimize it.

Data Sources

The names of the originating agencies for the map data are reported in the data sources section of the printed layout, the "About" page of the digital web map, or the corner of the digital web map. Include the agency and company names for each dataset, the name or a short description of the data, and the date of the data. This is a classic metadata element.

Best practices: The data sources element is very common and indeed, extremely useful for both the intended audience as well as for the future reference of the layout originator or project successor. Care needs to be taken that no abbreviations are used. For example, the layout designer may place something like this on the layout, "Data is from USGS EROS and NAIP 2006," which is scarcely better than not putting anything on there at all. While those may be common acronyms in the designer's office, the external map audience will certainly not understand them, the designer may forget what was meant a year from the printing, and the designer's successor on the project may not be acquainted with them. Furthermore, it is a common assumption that people who work in your close workgroup understand what the acronyms mean as well, and it would probably surprise most of us to learn that even they haven't deciphered these things! A better way of stating the above example is shown in Figure 3.17.

Style: The style guidelines for the data sources element are similar to all the other metadata text elements: keep it unobtrusive and minimal. However, the data sources have a slightly higher place than the disclaimer and network

High altitude aerial photo courtesy of the U.S. Geological Survey's Earth Resources Observation and Science data center

Elevation courtesy of the National Agriculture Imagery Program; 2006 imagery

FIGURE 3.17
An example of correct data source text: all acronyms except "US" are spelled out and no other abbreviations are used. This ensures that your map viewer (or even yourself several months down the line) can understand it.

path metadata elements as far as level of importance is concerned. Items to consider including about each dataset are as follows:

- Data dates
- Agency names
- Website
- A short description of how the data was used
- Potential limitations

Data Citations

Data that is contained on the map may require citations by the source agencies and authors of the data. Many times this is stated in a *Creative Commons* licensing agreement that goes something like, "you may use the data for any purpose as long as the source of the data is cited as follows: …" Therefore, your publication of the map, whether for in-house use or external use, needs to contain the citation in the manner specified. Sometimes, the exact string of text to be used is specified in the licensing agreement. Because this string of text may not be in a format that matches the surrounding text or is not appealing to the layout designer, the designer may have reason to be somewhat redundant in this case by citing the data source in the data sources element as well as citing it in a data citations element. An alternative is to ask the source author if your alternate wording is acceptable, and to get the permission in writing. Because the data citation is the originating author's way of receiving due credit for work that is being made freely available, it is important from an ethical as well as a legal standpoint to include it.

Best practices: If data with a citation clause is one of the major elements of your map or analysis, then you might consider making the data citation less of a fine-print metadata element and instead move it up in the element hierarchy of your layout. In other words, you will want to duly acknowledge the source of any data, without which you would not have been able to complete your work.

Logos

A logo is a graphical way of signifying a company name or brand and is used to enable rapid identification of what it is representing. Logos proclaiming

> The ability to throw the kitchen sink at one's output tempts many of us to clutter our maps. At the same time, we need not be constrained from employing capabilities not available to traditional cartographers.
>
> **Walker Willingham**
> *GIS Analyst, Earth Walker GIS*

authoring or sponsoring agencies are often displayed prominently on presentation maps and even on digital web maps as well. I am honestly not sure why logos are so ubiquitous on GIS layouts. Whatever the reason they are used, you can be sure that the logos will clash with the colors and style of the map product, and that they will either float unconvincingly in the margins or worse, in some prominent place on the map. Occasionally we can get away with it if we balance the logo with some other feature or tuck one black-and-white rectangular-shaped logo at the bottom or corner of the layout. The much preferred alternative to using a logo is to simply credit the authoring agency or agencies in a text box within the other margin elements. Often, though, the mapmaker does not have a choice and is directed by the higher-ups to have the logo placed on the product. In such a case, a gentle nudge toward the less intrusive method mentioned above, for the sake of design cohesiveness, is advised. Of course, people used to draw dragons on their maps, so ...

Best practices: If you are forced to include a company or agency logo on a layout where the brand is not of primary emphasis, then see if a black-and-white version of the logo can be used as this may interfere less with the overall color scheme of the layout as well as keep the eye from being drawn toward it. Also, sometimes a version of a logo that is a rectangle, square, or circle, rather than irregular, is available. If so, use that. As I alluded to earlier, a logo can sometimes be used to balance another element of the layout. For example, a circular logo can be placed opposite a circular north arrow for balance, or a square logo can be placed directly under a square descriptive text box.

Placement: The best place to put a logo, if it must be on the layout, is in the least obtrusive part of the layout, wherever that may be. Often this means the lower right-hand corner. Putting logos on either side of the title or at the beginning of a title adds to a cluttered feeling and leaves the eyes dancing around wondering where the most important piece of information is (hint: it should be the title or the map, not the logo!). But as with any rule of thumb there are some exceptions. One is if the map is being made for commercial use and the logo represents a brand as opposed to a company or agency. There may be a case for making this type of logo the primary element of the layout with the map element retaining only secondary emphasis. A skateboarding company, for example, that prints free maps for customers of great places to skate in town, might make its brand logo the most prominent part of the map since its purpose is primarily for marketing. However, for the analysts among us, we ought to keep the focus on the title and the map and forget the logo (see Figure 3.18).

Graphs

A graph shows the values of your data in diagram or chart form. Graphs can be any of several types, such as scatter plot, bar graph, pie chart, histogram, or bar chart. Many times an analysis map or even an informational map can benefit from the use of graphs to help highlight trends in the data.

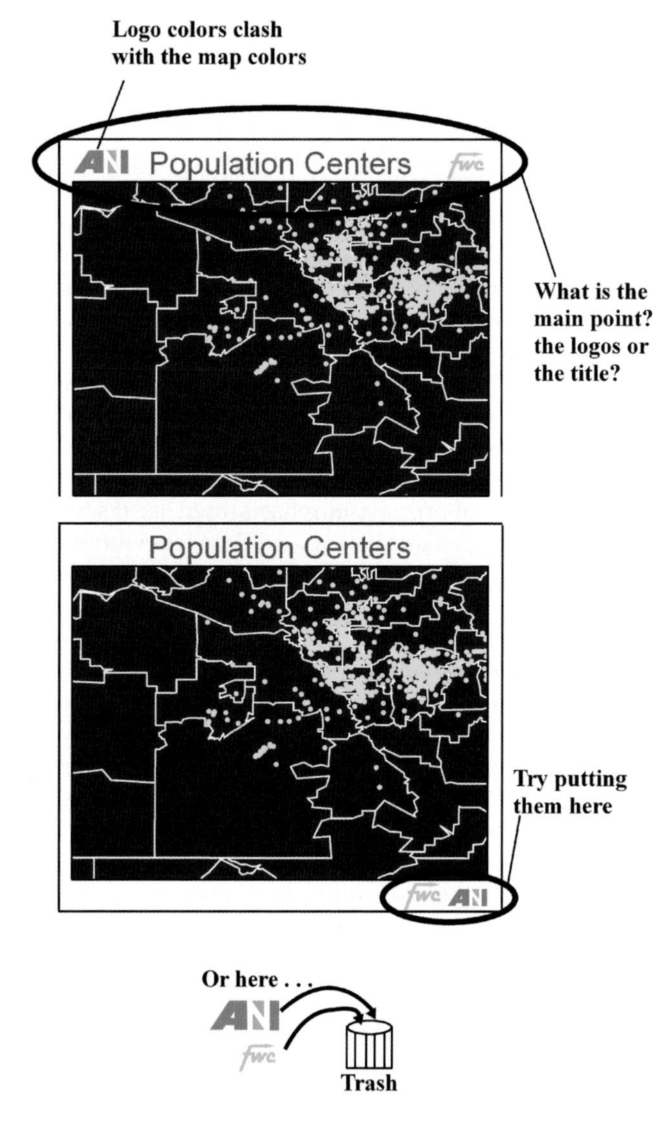

FIGURE 3.18
The map above is cluttered with logos. Deemphasize logos by relegating them to a corner or see if you can trash them completely.

For example, a layout with a gradient map showing dark colors for areas with high crime and light colors for areas with low crime might also contain a scatter plot with matching color schema showing crime rates on the x-axis and proximity to police stations on the y-axis.

Graphs can also provide ancillary data that, although not shown on the map, help to further the viewer's understanding of the material. For example, a layout with a map of population density by region might also contain several bar graphs illustrating the age distribution in each region.

Best practices: A graph ought to match the data shown in the map in terms of color and hue, if it displays the same data. As far as the type of graph goes, choose the type that depicts the trend or trends as simply as possible, and if you need to, insert a text box, arrow, or other highlighting mechanism that explicitly shows the viewer why that graph is important. Another method to achieve this is to provide a caption or headline for the graph. Pie charts, in particular, are often too showy due to distracting color choices and wind up hiding the ultimate trend rather than highlighting it. Indeed, this type of graph is often misleading if the pie slices number more than four because it becomes too difficult to determine what the individual values are without the aid of labels. A series of pie charts can be particularly egregious as they often don't contain the same categories or have them in different parts of each pie, making them difficult to compare.[3]

Placement: There are many possibilities for graph placement on a layout. A graph can float on top of the map element or it can be placed outside of the map element along with the margin elements. Much of this will be determined by how much white space (empty space) is contained on your layout, where the white space is, and how intimately the graph is tied to the map. If the graph has the same color scheme as the map element, you may consider placing it next to the legend or close to the map so the viewer can glance from the graph to the map or legend easily and thereby make comparisons on the fly. When the graph relates specifically to a single feature on the map, then it ought to have a leader line connecting it to that feature or some other appropriate means of letting the viewer know to which feature it is related. If you have a set of small, simple graphs that show additional details about discreet areas on the map (such as polygons), they can be positioned on top of the polygons, perhaps with connecting lines if needed.

Photographs

Digital photographs are used to enable the map viewer to see portions of the map in its real-world likeness. In some cases, the photographs are tied to a very specific location via coordinates and placed near where the corresponding location is on the map element. In digital, interactive maps, the photographs are often displayed in click events, but another handy method is to display related photographs in a side or bottom bar that can be scrolled, with photos that enlarge when clicked.

Potential photograph types for GIS layouts include the following:

- Pictures of the data collection process for presentation layouts such as a picture of a crew on a boat.
- Pictures that exemplify the data, such as pictures of representative habitat types for each habitat depicted on the map or pictures of the landscape being consumed by fire on a fire perimeter map.

- Pictures that enhance the map's usefulness, such as pictures of store fronts featured in a map of all the downtown coffee shops, so the map viewer can easily find them when on the street.
- Pictures of conditions such as drain-pipe pictures for each drain pipe in a quarter-mile area.

Best practices: The photograph should not be the focus of your layout. The best defense for including photographs on your layout at all is if they greatly enhance the viewer's understanding of the subject matter. For example, if you are mapping the distribution of squirrels across the United States, you may find adequate reason to include a photograph (or sketch, for that matter) of a representative squirrel for each of the ranges in order to help the viewer understand the difference in squirrel size across the mapped area. The photograph should be professionally composed and contain only relevant information in the foreground, if possible. The viewer should be able to look at the photo and immediately recognize the subject of the picture, rather than, say, wondering if the squirrel or the trees or the sky is the focus.

Placement: Photographs on print maps can be nestled anywhere it seems appropriate. An inappropriate location would be anywhere that it distracts from the title or map element. For example, if two photographs flank a title (i.e., photograph, TITLE, photograph) it causes confusion as to which element the audience ought to read first—the title or the photos? As stated earlier, digital interactive maps can be designed such that photographs don't appear until clicked, or if shown on the main map page, are displayed as thumbnails in a scrollable sidebar or floating element.

Style: If the photo is tied to a particular location on the map, then this needs to be identified in some way either by leader line(s), with a matching label on the map and photo, or through click events. Always include a caption with the photograph unless it is tied to the map with a leader line. For example, an overview picture could be captioned, "A view of Boulder County, Colorado, from the air." In most cases, strive to write the caption in a professionally worded, complete sentence format and remember to describe the photograph's relevance to the map subject. Do not, for example, caption a photograph of a boat crew with something that sounds like it comes from your high school yearbook such as, "Sammy and the crew hitting the waves."

Graphics

Graphics can include sketches, drawings, cartoons, illustrations, clip art, and other similar illustrative items. Whether displayed on a layout to provide information or whether they serve a purely decorative function, they are usually used sparingly or not at all on modern GIS layouts. Historic maps are distinguished by their sketches of dragons, gods, sea serpents, scrolls, dueling captains, and other superfluous graphics. In some special instances, these could be employed on modern cartographic outputs as well, especially if the map's purpose is more for entertainment than for research or reference.

Research or reference maps of particular genres may also use supplementary graphics in certain circumstances. For example, maps of archeological sites, burial grounds, and landscape designs are some modern map varieties for which these kinds of graphics could enhance, rather than detract from, the overall map goal. In fact, if you can recognize that your subject matter is sufficiently unique, you can use graphics to create a commensurately unique-looking map product. Putting that level of thinking into a map product rather than simply making a map that looks exactly like all the other GIS layouts is a hallmark of advanced design skills. If, however, your work genre is of the standard GIS analysis or informative nature, then the professional will generally steer very clear from any ornamentation, as our modern mode dictates.

Best practices: Do not use ornamental graphics unless your subject matter warrants it or you want to confront the modern fashion of information-only output.

Placement: Graphics can basically be placed anywhere and everywhere on the map layout from a light background image, to hundreds of sketches placed around the edges, to one illustration in the margin.

Style: Graphics can have any style and level of importance the designer deems appropriate. There is much leeway here and a lot of opportunities to make your map unique.

Map Number

A map number refers to the page number of a map that is contained within a printed map series. Map numbers are usually accompanied by a number signifying the total number of maps in the series. If your layout is part of a series, then it will be imperative that you include this information.

Placement: The map number element can be seen as similar in importance to the scale, author, and data source elements. It is usually placed in the margin along with those elements, but could also conceivably be placed higher in the element hierarchy if you wish to put it in large type in the upper-right or lower-left corner of the page. Adhering to the convention of placing it in one of those locations allows a researcher to flip through the series until the correct map number is located. Series-type maps usually also include a locator map. The map number and locator map are well suited for placement next to one another.

Style: Some examples of the way the map numbers are written include

- 1 of 10 *or* 1 of 10
- First in a series of ten
- Map number 1 of 10
- Series #1:10
- Map 1 of 10

Tables

A table is a series of data arranged in rows and columns, often with column headings. Because a GIS is composed of both the visualization of the features in their geographic space and the attributes of those features, the map output of the GIS can lose value when it shows only the visualization of the features and perhaps one attribute (such as the road lines and the road names) but not the other applicable attributes (such as road length, width, or condition). When the other attributes are lengthy or numerous, a layout designer may wish to include a table corresponding to the mapped features as part of the layout. This enables the viewer to connect the features with their attributes without needing complicated and sometimes impossibly numerous colors, symbols, and labels on top of the features themselves.

In some cases, we are faced with trying to decide between summarizing the attributes so that they can all fit onto the map or including the full data table on the layout. To show the distribution of cancer patients and cancer patient treatment costs across all counties in a state, you may be tempted to group the percentiles into four or five categories and color the counties with a corresponding gradient color scheme. However, the map could provide much more information if it is accompanied by a table showing the actual percentages for both attributes by county in descending order. There seems to be a misperception about the amount of detail that people can understand on a poster-sized, printed, presentation. For example, I have seen the advice, "don't insert long tables of numerical data or text" on conference poster guidelines. I think this advice comes from underestimating the human brain's ability to discern patterns in large amounts of data. If we format the table properly we can, indeed, present long tables, and we should if they further the message that we are trying to get across. This also increases the map's credibility.

Best practices: If you have a table showing many attributes for the same features, you could alternatively choose to create a series of maps with each one showing the features and one particular attribute. This is a judgment call based on the space available, the size of the features, and the complexity of the attributes.

Style: There are many table styles. The important guidelines are to ensure that the column and row headings are legible, that the data values line up, and that the column and row lines do not detract from the data. Because the map is still the most important element in the hierarchy, the table ought to include as few ornamental lines and borders as possible to ensure that it has less visual weight than the map. In fact, delineating every cell in a table with a border usually adds unnecessary clutter. Consider doing away with any lines at all and instead use white space as the main separator, with perhaps one or two simple neat lines separating the table from the elements above and below it.

To further enhance the table, you can add visual weight to certain elements. Anomalies in the data that may be too hard to detect on the map can

Fish Creek	25 | 150
Barberville	(340) | 180
Granton	50 | 200
Lakeland	62 | 250
Upwater	124 | 320

FIGURE 3.19
Tables can be formatted to clearly portray patterns in the data. In this example, the distance between the ratios and the place names increases in proportion to the denominators. When the data is presented like this it is easier to see that Barberville's numerator does not follow the pattern. The circle highlights this anomaly.

be highlighted in the table with a circle or bold text. These enhancements need to highlight only the pertinent data cell, not the whole table row. The conservative use of color—to differentiate assets (black) from debt (red), for example—can also add to a viewer's understanding of the data. Another trick is to change the locations of numbers based on some characteristic of the numbers. For example, a table of acres of impervious surface per watershed could show the impervious acreages at a distance from the watershed name in proportion to the watershed's size. This would be a great accompaniment to a map of watershed-level impervious surface normalized by watershed area because it would be presenting the viewer with the nonnormalized numbers in the table and normalized percentages in the map (see Figure 3.19).

Copyright

A map copyright states the author of the map layout and is sometimes accompanied by the date of copyright declaration. However, in the United States, all maps are automatically protected (except some government documents) and therefore explicitly stating the copyright is not an absolute necessity.[4] Creative Commons licensing is a newer method of licensing a creative work, like a map layout, that acts in addition to a copyright, and explicitly gives some rights to the licensee.[5]

Placement: The copyright information is usually left as an inconspicuous metadata-type element, and is given the least emphasis possible while still remaining legible.

Style: Usually, the text of the copyright includes the word *copyright* and the author's name or the copyright symbol (©) and author's name. The phrase "All Rights Reserved" and the copyright date(s) are optionally included as well. If using a Creative Commons license, the abbreviation (CC) plus the appropriate icon and/or text would need to be stated as advised by that license and your attorney.

COPYRIGHT TRAPS

A *copyright trap* or a *hook* is an error placed purposefully on a map so that if the same misinformation were to turn up on a rival's map, the original map owner would supposedly have definitive proof that the rival's map was directly copied from the original owner and simply resold for the rival's profit. Without this kind of misinformation, the rival could claim that it was simply coincidence that both maps display the same data. For example, the San Francisco Municipal Railway Map reportedly contained at least two fake streets (Geek Street and Moe Street) for this purpose.[6] However, whether or not these traps actually serve their purpose is unclear.[7] With the traps used as evidence, a few allegations of copyright infringement have surfaced in the United States and other country's courts in recent years. In one such US case, a US federal court found, "To treat 'false' facts interspersed among actual facts and represented as actual facts as fiction would mean that no one could ever reproduce or copy actual facts without risk of reproducing a false fact and thereby violating a copyright."[8] This statement reflects the concept that factual information cannot be copyrighted. Still, these traps may at least alert the author of a copyrighted map that their work was copied.

Projection

The projection of the main map element is sometimes reported in the layout margin of a printed map or in the "About" page of a digital, interactive map as part of the metadata for the layout. One purpose for providing projection information is as a means of alerting the viewer to the potential benefits and limitations of the map product. For example, knowing that a map is in a Lambert Conformal Conic projection would inform a knowledgeable viewer that the map is great for discerning the real-world shape of mapped features, but that using it for measuring the area of those features may yield erroneous results. However, if the viewer is not knowledgeable about projections, the same information could be described on the map in a more direct way by simply stating the limitations of the map (e.g., "Not for use as an area measuring tool").

Best practices: If your map audience is comprised of geographers, cartographers, or geoprofessionals, the projection information will be appreciated, depending on the subject of the map. However, even if the audience is comprised of only geoprofessionals, you may still want to consider laying out clearly what the projection consequences are as not all geoprofessionals are familiar on projection nuances. By including both forms of text, the projection and its limitations, you can ensure that everyone is adequately informed.

Placement: The map projection information is placed in the margin along with the other metadata elements that are similar such as the disclaimer, data sources, data citations, and copyright.

Inset Map

An inset map is a small map relative to the primary map, generally with a ratio of about one-eighth to one-sixteenth the size of the primary map element. There are two potential functions of the inset map. One is to show an area of the primary map in more detail by zooming in to a portion of the primary map and the other is to give an overview of the primary map's location by zooming out from the primary map. This latter type of inset is variously referred to as an inset, overview, or locator map. An example of the former—the zoomed-in type—would be a large-scale map of a particularly densely populated portion of a county. This inset would allow all of the mapped features to be viewed adequately. An example of the zoomed-out, overview type would be a watershed map that shows a stream in relation to the mainstem and other tributaries, shown in conjunction with a main map showing just the stream and detailed stream survey attributes.

Best practices: The same features and feature styles in the primary map element need to be duplicated in the inset map element. The orientation, or north position, should also be the same as in the primary map. Additionally, it is usually necessary to highlight the position of the inset's features on the primary map using a box, shaded area, or other such technique if it is of the zoomed-in variety. Lines coming out from the primary map's box and connecting with the inset map, called *rays* or *leader lines*, visually connect the inset with the corresponding inset box on the main map (see Figure 3.20).

Style: The inset map element should include its own scale bar, especially if it is a detail map. The overview style insets do not necessarily require a scale bar if enough spatial context is provided for the audience to orient themselves. An overview, or locator, inset map contains generalized data styled in

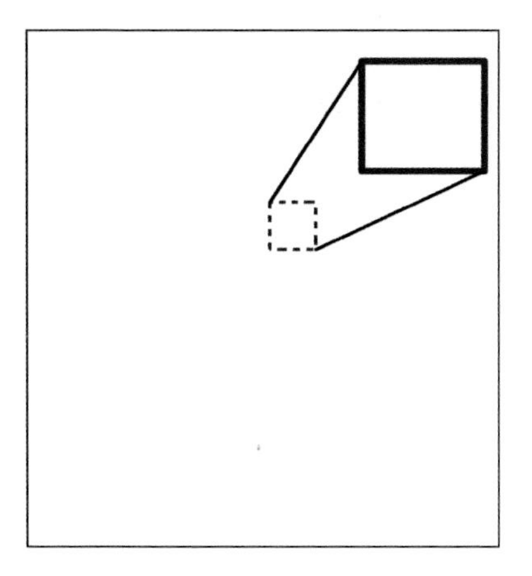

FIGURE 3.20
Inset maps that enlarge an area of the main map are tied to the main map in various ways. In this example, the inset map is connected to its corresponding portion of the main map via a bounding box and rays.

a subdued way. It indicates the location of the main map extent via an outline indicating the bounding outline of the main map or via shading.

Descriptive Text

Descriptive text refers to any text that furthers the map's purpose but doesn't fit into any of the above categories. Often, the specific nature of your map will require unique descriptive elements. A confident and skilled geoprofessional will recognize when and where a unique text element will enhance the viewer's understanding of the map. For example, a tourist map of ski trails might list websites to go to for current slope conditions.

Style: On a large poster-sized layout, remember to ensure the readability of all large blocks of text. This means that the font size must be large enough to read (14 point or 16 point at least) at a distance of one or two feet (see Chapter 4, "Fonts," for more text size guidelines). You should also double space the lines to make it even easier to read, look less dense, and be therefore less off-putting. Headings for large blocks of text should avoid the use of common terms like "introduction" or "study area" and instead use a description that is pertinent to the material such as "Tide Levels" or "The Great Lakes." This ensures that you do not waste one ounce of space on a word that is not specific to your subject. Consider a drop-cap for the first letter of the text block to draw the eye to that spot, or perhaps a special leading graphic like a simple icon.

Style

Once you have scrutinized the Layout Checklist and read through the accompanying element details and examples from this chapter, you will have a good idea as to what you will be putting on the layout. What should the overall style of the layout be, though? While the "Element Details and Examples" section did describe some style guidelines for each element, this section focuses on the general style of the overall layout.

To begin with, this section advises seeking inspiration from other maps, artwork, or websites to get an idea of what overall composition and style you like and would fit the subject matter of the map. Note that I say inspiration, not copying. Your data will lend your map a certain level of uniqueness, but as you start to add elements to the layout page, you will also find yourself moving them around in order to better fit your own style and the unique considerations that your subject matter requires. While inspiration may be sought initially, the final map product ought to avoid being too similar to the original inspirational work.[9] In its final form, the layout will be in harmony with the data of which it is comprised, its audience's expectations, and the most current cartographic styling.

Some things to take note of when perusing other maps, artwork, and websites are as follows:

- Overall feel: simple, complicated, scientific, humorous, medical, historical
- Colors: light or dark, patterns, color distribution around the page, background versus foreground hue, chart colors
- Element configurations
- Element separators such as neat lines, boxes, other graphics
- Font choices and styles
- Metadata text, style, and location on the layout
- Interactive components: what they do and how they work

Choosing favorite parts of various types of maps and creating an amalgam of them in your layout is another way to go. Yet another place to find inspiration for the overall design is to consider the context and audience where the map will be displayed. The following introduction to context explores this concept further, with specific regard to printed maps.

Context

Where will your printed map be viewed? Who will be viewing it? These two basic questions are the basis for most of the contextual considerations a mapmaker faces. Stemming from these are the secondary questions of how many people will be viewing the map, at what distance, with what surroundings, and so on. In the examples that follow, the contextual question of where a poster-sized map will be located forms the main design consideration.

Will your map be viewed as part of a poster gallery at a conference? A light-colored map with just the right "pop" in the important elements can provide soothing relief for bleary-eyed conference goers. Add a graph or two and your audience will enjoy at least pretending that they are scrutinizing the graph so that they may appear to be intellectual. If they happen to actually walk away having absorbed the simple message you were trying to convey with the map or graph, then you have served your purpose (see Figure 3.21).

Will this map be on a wall in the company hallway that happens to be painted a striking white color? Perhaps the opposite strategy could attract attention to your masterpiece. A great way to make a statement in this location is to use very dark, saturated colors to provide a pleasing contrast. Blues and grays are perfect for this, give a modern out-of-this-world feel, and are soothing all at the same time (see Figure 3.22).

FIGURE 3.21
One style approach might be to counteract the audio-visual chaos at conference poster events with a map that provides a modicum of calmness and sanity.

FIGURE 3.22
Dark colors provide drama for otherwise boring office walls.

How about creating a map for your workgroup that will be put on a gray cubicle wall within eye-shot of everyone walking by (including your boss)? Using candy-shop colors would brighten the gray cubicle wall and provide visual relief for those passing by (see Figure 3.23). Of course, an actual jar of candy sitting nearby could also serve your cause well!

Finally, consider your subject. Who is your audience? Tailoring the style to the audience's expectations can lead to a successfully received layout. For example, although you may not be in the medical field yourself, perhaps you are tasked with creating a poster of some health data for an upcoming medical conference. Instead of creating the layout in the same style as you have for other types of presentations, you can research the style that people in the medical field are accustomed to seeing. This might lead you to add in

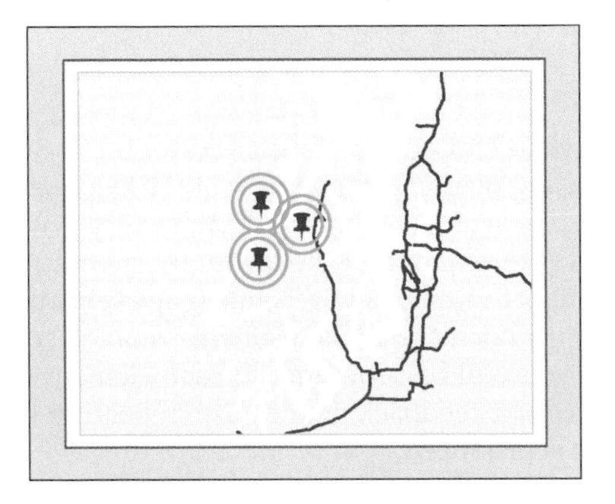

FIGURE 3.23
This map is designed with vibrant purple, green, and pink foreground features as a means of livening up a gray cubicle wall.

trend reports along the margins in the form of graphs and statistics tables. Another example is making a map for a business audience. If you aren't familiar with business style, then check out some of the leading business books and study their graphics. One of the things you would find is that the graphics are generally black-and-white line drawings with boxes, diagrams, and graphs. If you can get your layout to conform to a similar style, you might buy yourself some credibility from the get-go.

If you have a particular style in mind that is different from what your audience is used to seeing, it is still okay to use it if you have some good reasons for it. Some good reasons might include that you want to "update" the field with more modern-looking design, you found a better way, or you feel your audience is tired of the old way. Whenever introducing a drastically new design style to an audience, you need to preface your presentation with an explicit acknowledgment that what they are seeing is different from the norm, why you went ahead and designed it that way, and what the audience will gain from seeing it in the new way. This prevents your audience from simply concluding that you have no idea what you are doing.

The subject matter of your map may also influence its design. Consider archeology maps, which present some very interesting design options. Perhaps you are managing the GIS data for a site dig. When the overall site map is displayed on a large-format layout, you could take a cue from historical maps (after all, the site is historical) and place supporting information such as graphics around the main map. The graphics can fade out toward the main map in order to provide visual separation as well as to further the old-world look. The graphics can consist of pen-and-ink drawings like artist's renditions of what various parts of the site may have looked like in the past and drawings of what the site looks like now.

Arrangement

Once the layout style is chosen and the context for the layout considered, the placement, design, and overall arrangement of the layout elements should be easier. The first step in arrangement is to think in very general terms of what you want the viewer to see first on the layout by creating an emphasis map or, for a digital, interactive map, a wireframe (page schematic). The next step is to consider the arrangement of the elements more specifically, as well as the level of detail you will be providing in each element. The last step is to build the layout, ask for feedback, and revise, repeating these steps until a satisfactory layout is produced.

Before explaining the ins and outs of an emphasis map or wireframe, we need a short introduction to types of layouts, some typical compositions, and a warning. First, the type of layout that you are creating will greatly constrain your attempts at arrangement. For example, static report maps located in-line or full-page, website maps, and slide maps do not typically have a layout at all. At most, they will include a map, title, legend, and scale bar floating over the map, all potentially bound within a framing box. At the least they consist of the map itself with no supporting information, except as might be written in the accompanying text or caption, or as might be referenced in a verbal presentation.

Arrangement considerations are mostly applicable to larger format layouts, or complex interactive web maps. The following are some typical printed, poster-sized map layout compositions:

- One map comprising two-thirds of a layout; extensive margin elements
- Multiple maps in a time series; bounding boxes that group items; time labeled clearly

REPORTS WITH MAPS

If a map designed for an 8.5-inch by 11-inch report is not large enough to fill the entire space between the left and right margins of the report, consider using two maps, side by side, to avoid having too much white space on either side. Another trick is to create the map so that its size is half the width of the page minus the margin area, then place it in-line so that the text wraps around it. The result is more of a newspaper or magazine style layout and is more pleasing to the eye (see Figure 3.24). For example, if the page was divided into a table of six squares, two columns, and three rows, a map would ideally fit directly into one of the six squares. On an 8.5-inch by 11-inch page with 1.25-inch margins on the left and right, this would translate to a map approximately the size of a 3-inch by 3-inch square. Because of the small size, of course, the map should show only a large-scale view of the subject, or only a few features, or both.

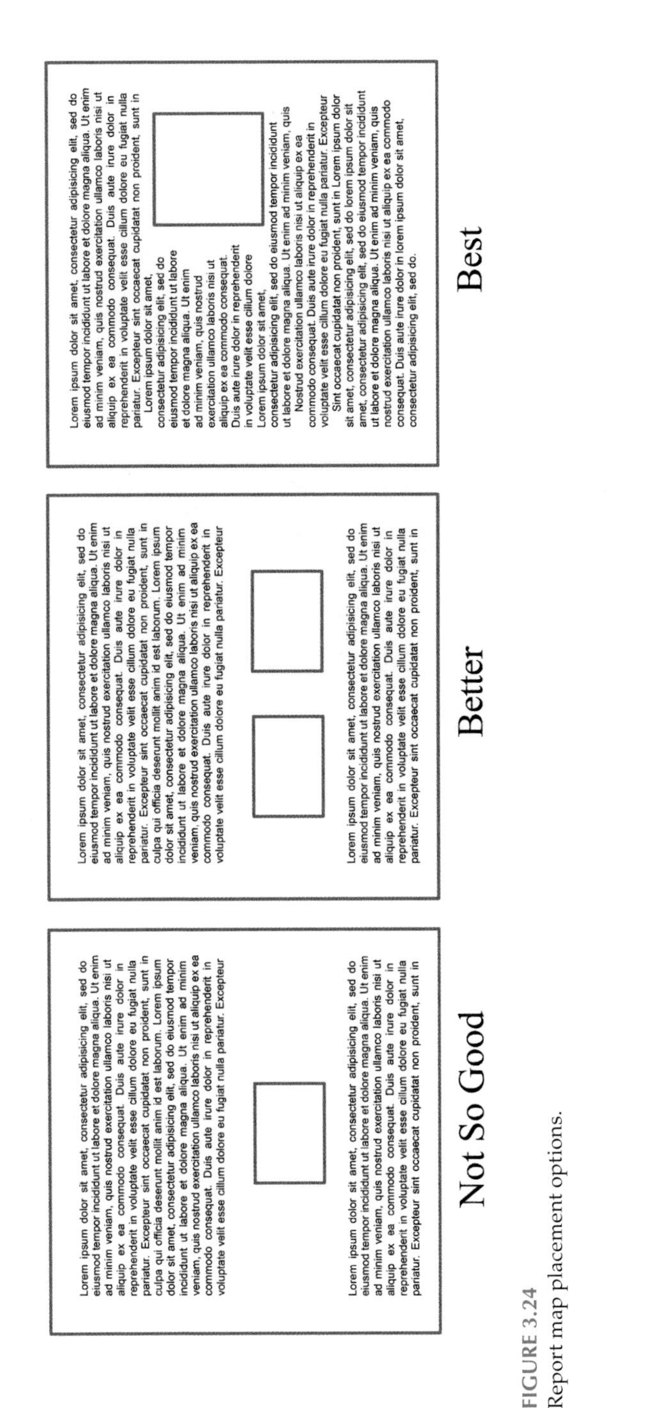

FIGURE 3.24
Report map placement options.

- Multiple maps of related data layers within the same geographic boundaries
- Working map with one large-scale map comprising the entire layout; specifications and legend information are placed in blank or nonessential spaces
- Conference poster with a small map relative to the poster size used for supporting information

These are some typical digital, interactive map layout compositions:

- One map comprising the entire layout with click events handling the interactivity
- One map with a title and a few ancillary elements handling the interactivity
- One map with many margin elements, all tied to the main map in various ways and with various interactive means
- One map with a data layer list located in the left-hand sidebar (very common, but not necessarily recommended)

And finally, a warning: while many concrete examples and guidelines are presented here (and other examples and guidelines can be found in other cartography texts) that will be useful to study for creative inspiration, the map designer ultimately needs to rely on a mixture of these guidelines, intuition, and previous experience to produce a layout that is appropriate to the map's unique data, subject, and shape.

Emphasis Maps and Wireframes

An *emphasis map* is a term borrowed from the world of web design that we apply here to paper map design. We'll use the term *wireframe* when applying the same concept to digital map design. Both terms are essentially interchangeable, however. They consist of an initial sketch of what your product will look like with specific attention to where the eye needs to be directed on an initial perusal of the document. In the case of a map document in which the mapmaker wants the viewer to read the title first and then look at the map, the sketch would show a title in large bold characters and then a map frame with a lighter outline and color scheme. All other elements would be sketched in a light and cursory fashion.

Make it a standard practice to put pen to paper and create an emphasis map, taking perhaps 10 minutes of your time, prior to embarking on any major map endeavor. It will force you to consider the main goals of your mapping from a design standpoint before your mind gets overwhelmed with the details.

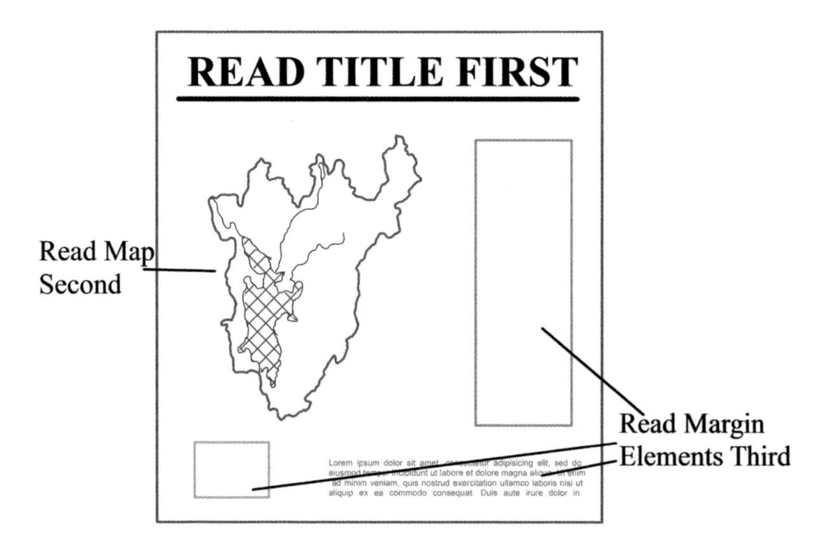

FIGURE 3.25
An example of an emphasis map. The largest and boldest elements signify where the mapmaker wants the map viewer to focus first.

As always, do not avoid taking the time to rework the emphasis map as you reconsider where the title will be and how much emphasis the map will be given, and so on. As an aside, the time you take, up front, to consider the layout will probably not decrease as you continue in this profession. Remember that there is a positive correlation between the quality of your layout products and the quality of your planning. Drawing the emphasis map or wireframe the old-fashioned way makes the process easy and painless. As a second step, however, it is helpful to translate the drawing into a digital format by using a graphics tool or a wireframe tool, especially when presenting the mock-ups to others.

You can see in the emphasis map in Figure 3.25 that the designer has already taken some preliminary steps toward laying out the elements, albeit in a generalized fashion. If at this point the designer did not have an idea of where the elements would be placed on the map, the emphasis map can instead be written in textual form such as: title = first, map = second, margin = third.

While it is common for a title to be created with the highest visual weight, it is conceivable that the mapmaker would want the primary map element to be the first element in the visual hierarchy, with the title being read only after a quick glance at the map. Other combinations are also possible but less common. Obviously, if a digital map will become an element in a website with an existing design, there is less room for page design and the time will be spent, instead, on ensuring that the map matches the surrounding frames as best as possible.

When a final emphasis map is drafted, the planning stage is complete and the process of placing elements onto the layout can officially begin. As you start to place the various elements onto the page, keep in mind how the overall feel of the layout is taking shape. If the layout is starting to look too cluttered, reassess whether the offending element is necessary, and if it is, try to find a way to minimize its visual impact. Although useless clutter is obviously not desirable, do not confuse that principle with the more insidious principle of keeping the map simple or "dumbed down." This is explained in the next section.

Simplicity versus Complexity

To begin with, the two output media types where simplicity is always okay are slides and printed, in-line report graphics. For these, aim for color continuity on all slides and report graphics for a polished look. If your map is not of either type, a certain amount of complexity is needed to justify the larger (or interactive) format. As long as your main data layers are appropriately highlighted, complex background data can add contextual understanding for the viewer that allows the map to be a rich resource of information. This leads to the following principle: do not be afraid to present extremely complex information as long as it is pertinent to your map's purpose. If it is presented in a coherent and unified manner, a complex map may even become your best, most cited, work.

Many times we confuse *readable* with *simple*. A map does not necessarily need to be simple to be readable as long as the appropriate highlighting of important information and organization of background elements are executed. An oft-cited reason for simplifying is the dumbing down of information for busy superiors and the nonexpert public. However, if you keep in mind the example of a standard US Geological Survey topographic map, which contains many details on trails, roads, rivers, towns, schools, and so on, in addition to innumerable contour lines, you can realize that most people with a high-school education are able to gain some understanding from them and feel a sense of knowledge gained as a result. That said, if your background data is both cluttering up the map and has no real utility, then by all means consider getting rid of it. Examples of background data that may, in some contexts, offer no additional value to the viewer while adding mindless clutter to the map are hillshade layers and small-scale aerial photos.

Overview inset maps, which give the viewer an overview of the general location for the main map, are elements that do require simplicity simply due to the lack of space allowed within their smaller map frames (this is a similar case to the in-line report maps mentioned earlier). When using an overview inset map, it is important to remember not to simply create a small-scale replica from the same exact data and symbology used in your main map. Detailed data presented in this manner need to be generalized first. For example, if development zoning data is presented in the main map, the zones will probably be too small in the inset map for the viewer to distinguish between them, and furthermore, they are probably not even necessary. The inset map for a development map might just show a simple box or single polygon that outlines the extent of the development zones with relation to nearby regional boundaries (which should themselves be generalized).

Those exceptions aside, the main point of keeping the complexity in your map is to empower your map reader. Let's say your boss asks you to create a customized map to be handed out to employees showing how to get from the office to the company picnic site. A typical dumbed-down approach would be to show the office, the picnic spot, and only the roads between that the driver will take to get there. What if the driver takes a wrong turn? In that case, the driver is out of luck and needs to get a detailed road map to get back on track. If you want to avoid disempowering the map reader but preserve the quick readability that comes from the original approach you could, instead, use a detailed roadmap as the main map and use a smaller inset showing the simplified map. Since the product would now resemble an old-fashioned roadmap, you could even fold it into 20 sections and hold a contest to see if anyone can refold it.

Many digital interactive maps made in the last decade contain a layer picker similar to a GIS software user interface. However, most users are only interested in using these maps for a single bit of information. To aid the user, then, it is much more reasonable to provide each thematic layer in its own web map, accessed via an overview page with thumbnails illustrating the web maps and their layer name.[10]

The last two principles to keep in mind as you start to add elements to the page are the design of the margin, if any, and the overall balance of the elements.

Margins

Margins, whether we're talking about the outer portions of the page on a paper map or the top and bottom menus and sidebars on a digital interactive map, must be integrated with the page and map seamlessly. And here the word *seamless* is used almost literally. Use white space as the "seam" between the margins and the map as opposed to thick and dark lines. Make sure the style—the fonts and the colors—has the same specifications as the map itself.

The convention in traditional architectural drawing is to place the margin elements, especially the key metadata, at the bottom or right-hand side of the map in a white rectangular area that spans the width or height of the page (see Figure 3.26). This ensures that flipping through a flat file drawer to find a particular map will produce the desired result instead of a heap of

FIGURE 3.26
Margin elements are normally placed at the bottom or right-hand side of the layout.

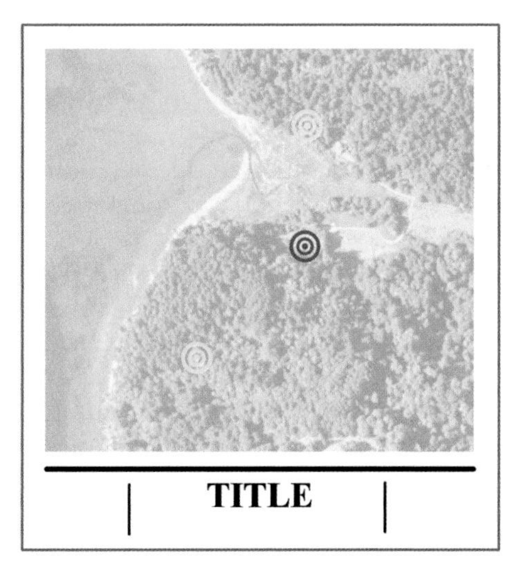

FIGURE 3.27
An architect's box at the bottom of the layout is elegantly split into three segments. The segments can be physical (straight lines or boxes) or elegantly implied (white space).

crumpled papers on the floor. This is one option to consider for paper map margin layouts.

This type of margin can be further split into discreet parts that focus on particular details of the map. For example, the layout in Figure 3.27 contains an architectural box on the bottom of the page, split into three segments, with the middle segment being the largest.

Floating margin elements, which some refer to as *integrated elements* are another option. With integrated elements, the map designer places margin information in the nonfocus areas of the map—the white space. This white space doesn't have to actually be white, of course, it just has to be a portion of the page that doesn't contain the important map information. This type of layout is an *infographic,* which is an area of design to study if you want to gain a deeper understanding of how this is done. Suffice it to say that infographics require a strong eye toward neatly aligning and grouping the margin elements so that they are easily recognizable and readable, but also not intruding on the map either physically or hierarchically. The rule of thumb is to use this method for stand-alone layouts and to reserve the architect's box–style for maps that are part of a series.

Balance

Balance is more than avoiding a lop-sided layout, although this is certainly a priority. Balance entails harmony in color, line weight, and content that ensures the layout is read as a cohesive whole rather than a sum of disparate parts. Practical application of the concept includes unifying fonts, counter-weighting peninsular geographic features with other elements, and correctly utilizing empty space. (Remember your grade-school art teacher's advice to make your drawing fit the whole page? Risking a stern reprimand from said teacher, some white space can be defensible on a printed map.)

White space used as a visual separator between elements needs to be balanced throughout the layout in terms of spacing and alignment. This is best explained via the example in Figure 3.28, which is a close-up of the lower right-hand portion of the map in Figure 3.27 of the "Margins" section.

Content balance in terms of style is also a consideration. As the style of your finished map ought to be chosen at this point, you can choose appropriate fonts, color choices, and element arrangements that will conform to that style. For example, different fonts can portray different styles (see the examples in Figure 3.29). See Chapter 4, "Fonts," for more information on choosing the right fonts.

Balancing out graphics by shape and color is also advisable. For example, counter-weight a large circular north arrow with a logo by placing it on the opposite side of the page. Taking this one step further, the logo could be displayed in a black-and-white color scheme to match the black-and-white north arrow, with the added benefit of deemphasizing the logo.

Modern map layouts should contain elements of both symmetry and asymmetry. The main parts of the layout may be asymmetrical along the horizontal axis, for example, while the subsections remain symmetrical around an imaginary vertical axis. Conversely, the asymmetry may be around the vertical axis while the symmetry is present around the horizontal axis (see Figure 3.30).

How will you decide what kind of symmetry is best for your map? As with all of the other aspects of the arrangement phase of the mapping method, experimentation, coupled with inspiration with an eye toward standard practices, is all that is necessary to achieve a balanced design. As the Staples button says, "that was easy."

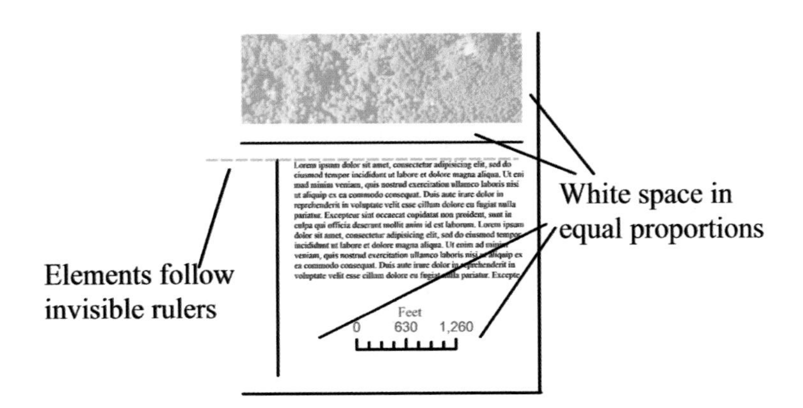

FIGURE 3.28
All white space needs to be carefully aligned so that it is in equal proportion around the layout elements. This is accomplished via on-screen rulers or guides and a lot of patience. Always check to make sure the elements are aligned properly at a 1:1 scale before printing.

Historic

FIGURE 3.29
Different fonts evoke different styles.

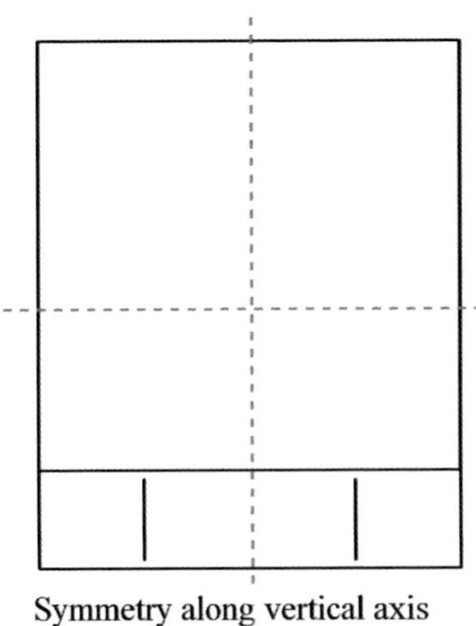

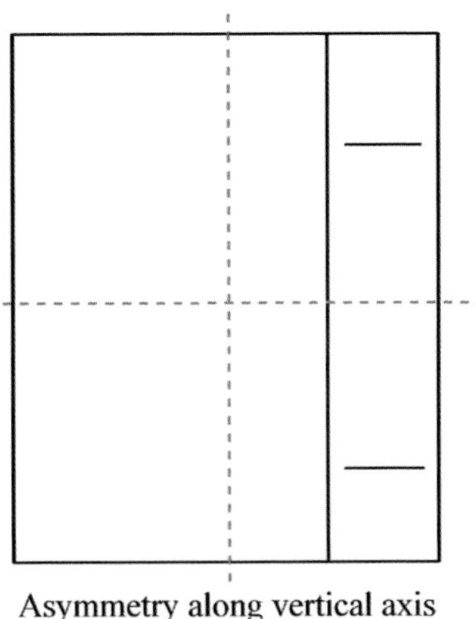

FIGURE 3.30
You can put your layout elements on the page in a symmetric fashion along the x- or y-axis.

Endnotes

1. As reported by Cheryl Lu-Lien Tan in "Prepping Food for the Eyes," *Wall Street Journal*, August 25, 2011.
2. Edward Tufte, *Beautiful Evidence*, Cheshire, Conn.: Graphics Press LLC, 2006, 133.
3. More information on the use of pie charts, including when they are and are not appropriate, can be found in Stephen Few's online report: S. Few, "Save the Pies for Dessert," *Perceptual Edge Visual Business Intelligence Newsletter* (August 2007), http://www.perceptualedge.com/library.php (accessed October 7, 2013).
4. See Title 17, U.S.C. § 101, et seq. This information is provided for informational purposes only and does not constitute legal advice.
5. More information about Creative Commons licensing can be found at http://www.creativecommons.org.
6. C. Squatriglia, "Mapmakers' Sleight of Hand: Cartographers put 'Bunnies' on the Map, Tricking Copycats, Sometimes Tourists." *San Francisco Chronicle*, August 12, 2001.
7. Mark Monmonier, *How to Lie with Maps*, 2nd ed., Chicago: University of Chicago Press, 1996.
8. *Nester's Map & Guide Corp. v. Hagstrom Map Co.*, 796 F. Supp. 729 (E.D.N.Y. 1992).
9. Map products are protected under copyright laws in the United States.
10. See Brian Timoney's blog post: B. Timoney, "Why Map Portals Don't Work," *MapBrief* (February 2013), http://mapbrief.com/2013/02/05/why-map-portals-dont-work-part-i/ (accessed October 7, 2013).

Suggestions for Further Reading

Cairo, Alberto. *The Functional Art: An Introduction to Information Graphics and Visualization (Voices That Matter)*. Berkeley, CA: New Riders, 2012.

Few, Stephen. *Information Dashboard Design*. Sebastopol, CA: O'Reilly Media, Inc., 2006.

Robbins, Naomi. *Creating More Effective Graphs*. Hoboken, NJ: John Wiley & Sons, Inc., 2005.

Yau, Nathan. *Visualize This: The FlowingData Guide to Design, Visualization, and Statistics*. Indianapolis, IN: Wiley Publishing, Inc., 2011.

Study Questions

1. What are the characteristics of a good map title? Give an example of a good map title.
2. What is chromatic contrast?

3. What are graticules? Why would a cartographer include them on a map? Provide at least two reasons.
4. What are the two types of inset maps? Describe each in a sentence or two.
5. What is a copyright trap? Would you ever include a copyright trap on a map? Why or why not?
6. What problems can arise when including logos on the map page?
7. How should dataset sources be cited? In short form (acronyms) or long form?
8. How does a cartographer create a balanced map layout?
9. Under what situation would you not want to include a scale bar on the layout?
10. What is an emphasis map and at what point in the design process would you draw one?

Exercises

1. Tables are difficult to integrate onto a map layout. Find an example of a table that is well integrated with a map on a layout. You may find it helpful to search using the terms *infographic* and *map* to find one. Briefly describe it in terms of font style, number alignment, grid style, and location on the page. Remark on what you like about its design aesthetic.
2. Should maps be as simple as possible, extremely complex, or somewhere between the two? Find one example of each of these, and write a paragraph for each, discussing how the level of complexity works or doesn't work for the map. Include a screenshot of each map in this short essay. This is a subjective exercise and as such student's ultimate conclusions may vary.
3. Take a look at the Natural Earth data download site at http://www.naturalearthdata.com. Choose one of the datasets (e.g., 1:50m Admin 0 – Countries) and create a data citation for it including the dataset name, download site, and date. You may want to view the Terms of Use on the Natural Earth About page for help.

4

Fonts

> Some of the most important aspects of cartography are those which are most subtle like colors and fonts.
>
> **Matthew Gilmore, IT Specialist**
> *District of Columbia Department of Consumer and Regulatory Affairs*

Rise above nascent professional mediocrity—with font theory! You need to know now—right up front—that getting the fonts right makes your map look like serious business. It's easy to get so caught up in the symbolization of data that the labeling of that data gets relegated to "that which you do at the end of the project when there isn't much time left," which is a travesty because being satisfied with placing text on the map and layout in its default form or, at most, choosing between Times New Roman and Arial, is the best way to a mediocre map.

Creating professional and one-of-a-kind maps and layouts requires the study of font theory and the techniques for modifying fonts. Some of us learned font theory the hard way: by picking up bits and pieces of information here and there over the course of many years. After all, it does seem like it would be a waste of a geoprofessional's time—especially a geoanalyst's time—to wade through the minutia of font theory to look for and learn the parts that apply to cartography. The aim with this chapter is to provide enough font theory to lift your map labels out of the realm of the novice, and it does so with particular attention to the parts of font theory that are relevant to maps. This gives you the necessary information to make great maps without wasting your time.

The first things to learn are the various font categories, what each is useful for, and some basics on the most common fonts available. After that you need to learn about various options to modify fonts in order to achieve certain effects like emphasis, de-emphasis, visual continuity, and so on. Once you choose a font and modify it as needed, the third concept to learn is text placement strategy. Lastly, a list of resources is included for those who want to dabble in nonstandard fonts or perhaps even create your own.

Choosing the Right Font

I must note here that when I use the word *font* I mean a particular design of letters with a name like Futura or Arial. Although *font* is rapidly becoming the word that everyone uses to identify such letter styles, the more correct term is actually *typeface*. The *typeface* is traditionally considered the name of the letter style group such as Futura, while the font is traditionally considered to be a particular version of the typeface like Futura 12-point italics. More specifically, the font is supposed to be the directions that the computer uses to define the way the letters look on-screen or on a printout. However, these days most people do not refer to it in that way.

The font you choose for various elements of your layout, including map-element feature labels, is influenced by the style of the layout, the feature that is being labeled, the function of the text, and several other parameters. Fonts are differentiated by their style, which is comprised of the character of the font along with its intrinsic letter height, width, and line thickness. The style of each font, even within the same category, can vary enormously and therefore it will be beneficial to keep in mind what each of the major fonts looks like as you go about designing your layout and map. Your font choice is an aspect of the map that can easily add a touch of differentiation between your map or layout and those of other mapmakers, thus adding to your image as a professional rather than looking as if you simply pick all the defaults.

The font categories discussed in this chapter are serif, sans serif, decorative, and script. While it is true that decorative and script fonts can be put into serif and sans serif categories, they are in their own sections here because they are aesthetically very different from the more common serif and sans serif fonts.

Serif versus Sans Serif

It's good to know the difference between serif fonts and sans-serif fonts because your understanding of them will affect your choice of font for different map purposes. (There is some debate over whether the differences between these two font groups actually affect the reader in the ways discussed here, but more on that later.) First, let's just figure out what a serif font is and what a sans-serif font is. Simply put, a serif font contains what I flippantly refer to as "doohickeys" while a sans-serif font does not (see Figure 4.1). If you prefer a more scholarly explanation, *sans* comes from the Latin term *sine* (without), and *serif* (short lines), which originates most likely from the Dutch word *schreef* (stroke), or from Middle Dutch *schriven* (write), or from Latin *scribere* (write). Fonts will usually fall into either of these two categories.

Keep in mind that even within the same category, the fonts can still vary quite a bit. It can be useful to keep a sample of the lettering for each of the major fonts in front of you while you are choosing the font for your map and layouts. The sample in Figure 4.2 can get you started and also illustrates the many varying intrinsic characteristics (explained in the next section) of each

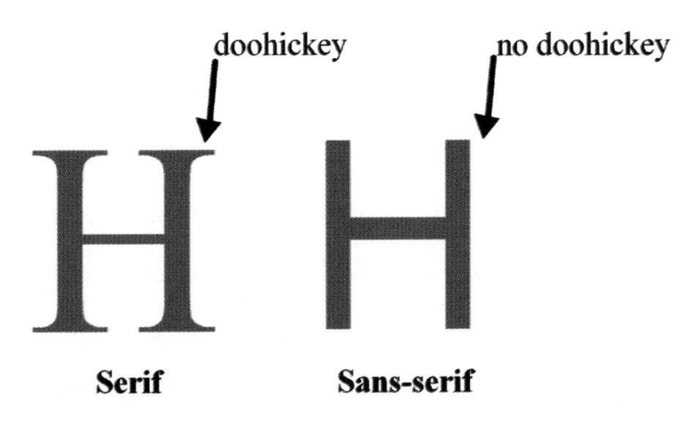

FIGURE 4.1
The presence or absence of "doohickeys" is the differentiating factor between serif fonts and sans-serif fonts.

font. The fonts are shown in 12 point so you can see that there is a difference in line weight, height of the letters, and style of lettering in each. The list contains Windows fonts, but when there is a Mac equivalent it is listed in parentheses. Many more fonts are available on your operating system; you can purchase fonts and download shareware and freeware fonts; and you can reference open-source fonts via the code for interactive web maps.

The presence or absence of the doohickeys relates to the readability of text under certain conditions like distance and resolution. In terms of legibility, where legibility refers to the level of distinction between characters in a font, the serif fonts do a better job at close distances on high resolution outputs such as a paper printout. This difference in legibility is best illustrated by example.

In Figure 4.3, the first three letters all look the same in the sans-serif version of the word *Illustration*. This would lead you to the correct conclusion that a serif font is the appropriate choice for blocks of text containing more than one sentence that are intended to be read at close range. It used to be that web maps used sans-serif fonts exclusively, due to low-resolution monitor restrictions. Most digital devices today have much higher resolutions than they used to, so we now see both sans-serif and serif fonts and they both look good. If you are creating a web map you can choose which you like best.

For shorter text blocks, like a title on a poster, you should consider using a sans-serif font because it truly is more readable at a distance despite the handicap previously mentioned. I have to admit that when I first learned this "rule," I didn't buy the argument and wondered if choosing a sans-serif font for long-range viewing optimization would really make much of a difference. So I tested it by printing out the same word at 100 point, one in Times New Roman and the other in Arial. Sure enough, at a distance of six feet I thought the Arial was much more readable than the Times New Roman. Try it yourself!

Serif Fonts	
Baskerville*	Mapping
Bodoni MT	Mapping
Bookman Old Style	Mapping
Courier New (Courier)	Mapping
Garamond	Mapping
Georgia	Mapping
Palatino Linotype (Book Antiqua)	Mapping
Times New Roman (Times)	Mapping

Sans-serif Fonts	
Arial (Helvetica)	Mapping
Century Gothic	Mapping
Comic Sans MS	Mapping
Gill Sans MT	Mapping
Impact	Mapping
Lucida Sans Unicode (Lucida Grande)	Mapping
Tahoma (Geneva)	Mapping
Trebuchet MS	Mapping
Verdana	Mapping

FIGURE 4.2
A few common fonts are listed here with the font name on the left and the word *Mapping* on the right, printed in the font named. Use this as a resource when deciding which font to use, or make one of your own.

Illustration	Illustration
Serif (Times New Roman)	Sans-serif (Arial)

FIGURE 4.3
The word *Illustration* is shown here in a serif font and a sans-serif font.

The Georgia font (serif) and the Verdana font (sans serif) were specifically created for digital display. They were designed by Matthew Carter of Carter & Cone Type Inc., for Microsoft in 1993 and 1996, respectively. Because monitors display images by way of small squares of light, or pixels, fonts can appear jagged on lower-resolution devices. Georgia and Verdana, however, don't have these issues, so they are particularly well suited for any web maps that you might create.

Verdana, which got its name from a combination of the company's verdant (green) surroundings and the daughter (Ana) of the project manager, is highly readable on a computer screen even at small sizes.[1] Wide letters and letter spacing and tall lowercase letters help increase its legibility. Individual letters were designed so they would be different enough to tell apart. In particular, the capital "I" is differentiated adequately from the lower case "l," unlike many other sans-serif fonts, by the use of horizontal bars running perpendicular to the vertical bar, just like in most serif fonts. Other letters are designed with the same letter-pairing limitations in mind. Letter pairs such as "ff" and "fi," for example, do not overlap.

Georgia has become an enormously popular font over the last decade because it is elegant and very easy to read. Interestingly, it got its name from a tabloid headline that suggested aliens had landed in Georgia. Similar to Verdana, it is legible at small sizes and has tall lowercase letters. It edges out Times New Roman for onscreen readability and has attractive italics. One caveat with the Georgia font, however, is that its numerals don't always look nice on map labels because they are offset from one another and therefore don't "sit" on the same plane (see Figure 4.4).

The numerals are considered "old-style" because they read more like regular text; that is, some of the numbers are higher and some are lower. Because of that, some consider them easier to read than numbers that are all the same height (this is a similar argument to the one that says lowercase letters are easier to read than words in all uppercase). As I mentioned before, map labels may not look their most attractive in this font if the labels consist entirely of numbers. If you are labeling spot heights on a web map, for example, you may want to use Verdana as its numerals are all of equal height. If you just have a number here and there inside a large text block, however, Georgia numerals are perfectly appropriate.

123456789

FIGURE 4.4
Numerals written in the Georgia font are staggered.

Decorative

While the serif and sans-serif fonts are going to be the major tools in the geoprofessional's arsenal, there are a couple of other font types of which you may want to be aware. The decorative font type (also called a *display font*) is a style that is reminiscent of a particular time in history, a place, a people, or some other style category. Decorative fonts might allude to such things as medieval knights, Celtic history, space, science fiction, and so on. In a geographic information system (GIS) map or layout they are reserved for use when the feelings they evoke directly match the subject of the map, which is to say not often for the majority of us in the analytical field. Analytical maps call for the standard font choices in either serif or sans serif. Decorative fonts convey the opposite of the analytical approach we want with those types of maps.

Perhaps though (stay with me here) you will someday decide to create a map showing all of the real towns that a fictional cowboy movie character visited in some old Western movies. This, obviously, is not your typical analytical GIS map. Here's the perfect place to use that Wild West font you downloaded from a freeware site long ago but never had occasion to use! It might be perfect for the title and subtitle of the map. However, be warned that your use of the font needs to be limited to those short, main features—like the title and subtitle—because decorative fonts in long blocks of text tend to overwhelm a map viewer. In fact, a font that is not immediately familiar to a reader will significantly slow that reader's comprehension rate.[2] Fonts that are decorative in nature, such as those shown in Figure 4.5,[3] can often be found for purchase, shareware, or freeware online; see the "Resources" section of this chapter for details.

Script

Script fonts (also called handwriting fonts) are also somewhat decorative in style. As with the decorative fonts, they are hard to read in large quantity and so should be reserved for small bits of text and only when appropriate. Script fonts contain extra flourishes, such as curls at the ends of letters, or mimic calligraphy or handwritten cursive and generally lend a more elegant, formal feeling to the text than a simple serif or sans-serif font.

Script fonts are somewhat more useful than decorative fonts for the GIS practitioner in that they are commonly used for labeling bodies of water such as bays, oceans, straits, and rivers. They could also be employed as a

FIGURE 4.5
These are three examples of decorative-type fonts.[3]

label font for maps with a lot of feature labels requiring differentiation. And, as in the Wild West example previously discussed, there could be other circumstances that would warrant a script font in the title or small text boxes of a layout design. Although this is an equally far-fetched example for a typical GIS analyst, I could imagine a case where a geoprofessional is asked to plot all of the public marriage ceremony locations in a city (such as parks, beaches, and so on). Such a map would be an ideal candidate for the use of a script font on the point labels as well as the title and subtitle considering that script is a traditional font choice for wedding ceremony correspondence.

You can discover several script fonts on your operating system already, and you can also find them online. Some examples are shown in Figure 4.6.[4]

Letter Height, Width, Line Thickness

The *intrinsic* letter height, width, and line thickness is built into the font that you choose. For example, the Bookman Old Style font has taller, wider, and thicker letters than the Garamond font. This holds true for uppercase letters, tall letters with ascenders (like h and k), short letters (like a and e), and below-line letters with descenders (like g and j). See Figure 4.7.

Similarly, the Arial font and the Times New Roman font differ in their letter heights to such a degree that a 100-point Times New Roman font is roughly equal to a 96-point Arial font. For this reason, always consider your font choice when deciding on a point size and especially when revising a map. Changing the font during a map revision may require a subsequent change in font size to achieve the same readability as before. For example, if you have city labels in 8-point Arial and you change them to Times New Roman but keep the same 8-point size, you may find that the labels are just too small

The Freestyle Script

An elegant script called Exmouth

An elegant script called CommScriptTT

FIGURE 4.6
These are three examples of script-type fonts.[4]

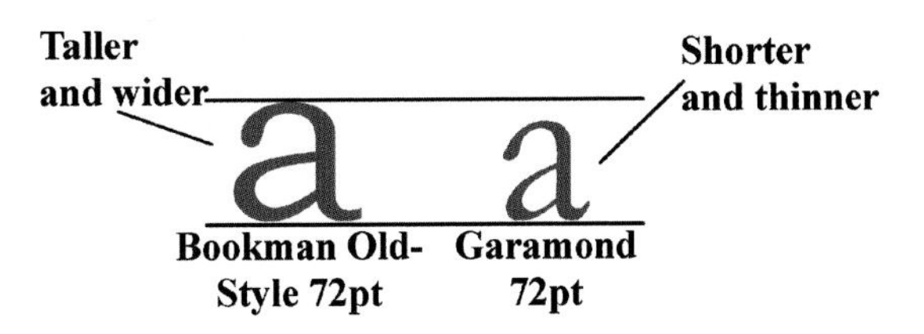

FIGURE 4.7
These two fonts show a wide degree of variation in height and width despite being the same point size.

is	is
Times New Roman	Courier New

FIGURE 4.8
The Courier New font, shown here next to the more common Times New Roman font, is a monotype font, meaning that all the letters are of a fixed width.

433509	433509
256798	256798
613257	613257
659310	659310
Times New Roman	Courier New

FIGURE 4.9
Blocks of numbers are easier to read in monotype fonts, as this example of the monotype font, Courier New, shows. (You might notice that the Courier New font is used in Chapter 5, "Color," for examples that contain stacked numbers.)

to read. If you want to keep the Times New Roman font you will have to increase the point size to 9 or 10 to achieve the same readability.

In terms of font width, most fonts contain letters with proportional widths but a few do contain fixed-width letters. These are referred to as *monotypes* and they can be either serif or sans-serif types. The Courier font is one such monotype in a serif style (see Figure 4.8).

Although Courier is often thought of as boring and perhaps a bit ugly, maybe because it reminds one of an old typewriter document, it can be useful to the geoprofessional when tables of numbers are displayed on a layout. Because its numerals are all the same width and have a nice amount of white space between them, it is much more readable than many other font number sets (see Figure 4.9).

The weight of the characters is another variable to consider when choosing a font. Line weight, or the thickness of the letters, varies by font and will create a different level of impact on your reader depending on which you choose. One example of this is shown in Figure 4.9, where you can see that Times New Roman has a much heavier weight than Courier New.

Modifying the Font

So you've seen how the height, width, and weight of fonts can vary somewhat substantially between fonts without any manipulation on your part. However, starting with those base characteristics, you will then need to increase or decrease the font's readability and impact yourself to fit your map's unique circumstances. The first, most obvious way to accomplish that is to change the point size. After that, there are several other techniques, such as character spacing, italicizing, and color modification that you may decide to use. First, let's talk about point size, and then we'll take a quick look at the other options.

Point Size

While we've already seen that there is some intrinsic variation in the font sizes, we also need to be aware of general point size guidelines in relation to the amount of emphasis needed for a particular block of text and the viewing distance. For example, a letter-sized printout intended for close-range reading would be best served with a 10-point to 14-point font for normal text and a 14-point to 20-point font for emphasis text like a title. Remember that 1 inch, or 25.4 mm, is roughly equal to 72 points, where the term *point* refers to the desktop publishing (DTP) point system of measurement.

$$1 \text{ inch} = 25.4 \text{ mm} = 72 \text{ points}$$

Furthermore, the 1 inch we refer to is the total distance between the height of the tallest letter (like and H or I) and the lowest point of the below-line letters (like g or y); see Figure 4.10. This conversion depends somewhat on the font you have chosen, of course.

Since viewing distance is a major variable in determining point size, let's focus on two major viewing distances. The first, as in the previous example, is a close-range viewing distance of about half a meter. Note that we are talking about point size in relation to viewing distance, *not* the layout size. Why do this? Perhaps it is best explained with this simple example: a small map in a report should have the same font sizes as a large map in a report because both will be read from the same distance. Now, for this viewing distance we'll eliminate the possibility of using a font size of less than 6 point because it would not be readable at any distance. Optimally, a 12-point to 14-point font for main text and a 16-point to 20-point font for title text should be used. Some variation around these is also acceptable given certain conditions or depending on your font choice.

FIGURE 4.10
Point sizes are related to the height of the font as measured from the bottom of the lowest character to the top of the highest character. Not all fonts of the same point size will have the same height, though, as slight differences in character height exists between fonts.

Poster-sized printouts for a conference poster session or for display on an office bulletin board are another matter because they are typically viewed from afar. But how far will this viewing distance actually be? If you were to search on the Internet for typical poster viewing distances you would find most people recommend that the poster be readable from a "typical" viewing distance of 2 to 3 meters. Have you ever seen pictures of people looking at posters in a poster session? Or have you observed poster session attendees actually reading posters? Well I have, and my conclusion is that *most* people are going to read your poster from an arms-length distance, or no more than 1 meter away. Now, either these people are viewing posters at a much closer distance than people realize because the poster presenters did not follow the text size guidelines and make their text large enough to read *or* it is simply a more natural distance at which to read. Perhaps it is a more natural distance because if you stand too far away from a poster you feel as if you are hogging the aisle space or people start to walk between you and the poster, or worse, someone comes up and stands in front of you to read the poster herself.

So, do I think this means you should use a smaller point size for your poster text than what everyone recommends? Yes and no. It means that if you want to, you probably can use a smaller font size for the body text than is normally recommended. But another consideration is at play here, and it specifically has to do with posters, and that is the issue of time. Considering that a large-format poster usually displays a brief overview of a topic, people will not want to linger over it, and a larger text size can suit that need for a speed-read. Whatever font size you use for the poster's body text and labels, you do want to be mindful of using larger text sizes for the title and subtitle to accommodate a viewing distance of more than 1 meter. That's because those titles are what draw people in closer to your poster.

A large-format technical map is the same size as a conference poster, but has completely opposite goals. It's definitely meant to be read at close range and at long length. Its reason for being large is simply because all of the information that is needed by the technical crew is impossible to display all at once on a smaller sheet. The body text, which on a technical poster will be confined to labels, legends, and other explanatory information, can be as small as you might find on a letter-sized sheet. The title and subtitle information can also be as small as titles on letter-sized sheets because these maps are meant to be deciphered and examined in minute detail, not quickly glanced over at a 1-meter distance.

So what are some specific guidelines for point sizes at specific viewing distances, you might ask. I wondered this too, and while I did find many recommendations for various font sizes, I found that they all varied somewhat and I wasn't sure if they had ever been adequately tested. So, after countless hours printing out different fonts at different point sizes, taping up the fonts on a wall, and standing at measured intervals from the wall, I came up with some guidelines that, I believe, are more realistic for our printed GIS maps and layouts.[5]

By testing the readability of various fonts at various sizes and distances, I have formulated charts to help guide you when designing your layout text.

Two of the most common fonts, Times New Roman (serif) and Arial (sans serif), are represented in these charts along with viewing distances and a few text types. My research shows that the difference in intrinsic font size is not going to make much of a difference in terms of which point size you choose until you get to at least a viewing distance of 3 meters. At that distance, the smaller font, Times New Roman, will need to have a slightly larger point size to get the same readability as Arial. Figures 4.11, 4.12, and 4.13 show the results of the testing that I conducted on readability. You can use these as a guide for your own work, but keep in mind that these are subjective, which is why I provide such large ranges in point size recommendations. If you are unsure of which point size to use, just print out several options in your chosen font, tape them to the wall, and stand back at your anticipated viewing distance.

Using these graphs, the guidelines for a poster that is truly meant for long-range viewing would be 18-point minimum for body text, 90- to 120-point

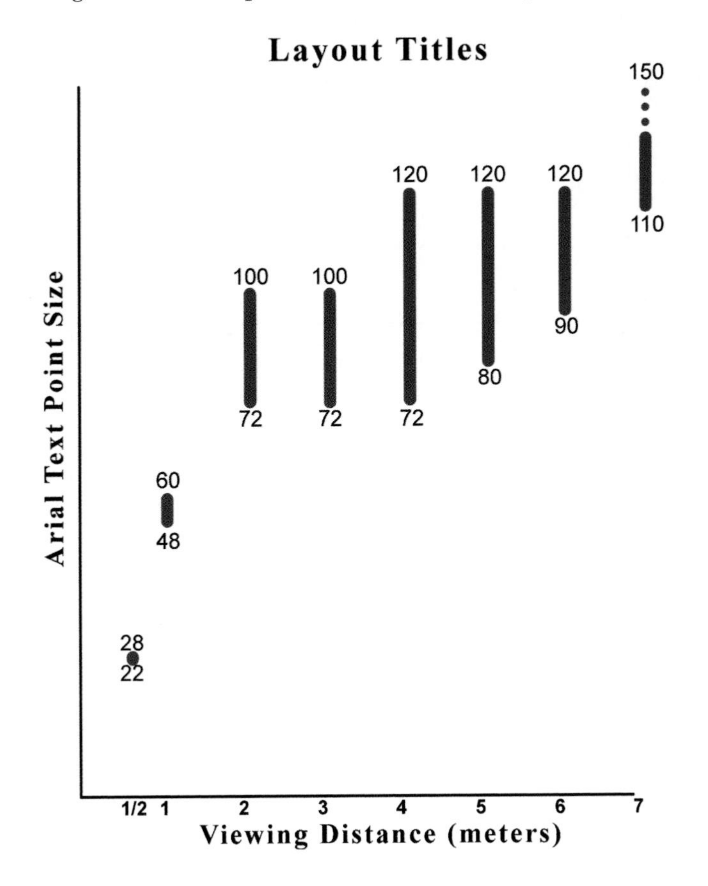

FIGURE 4.11
Approximate point sizes to use for map titles depending on the expected viewing distance. A map intended to be read at a distance of 3 meters, then, should have a title in the range of 72 point to 100 point if using the Arial font.

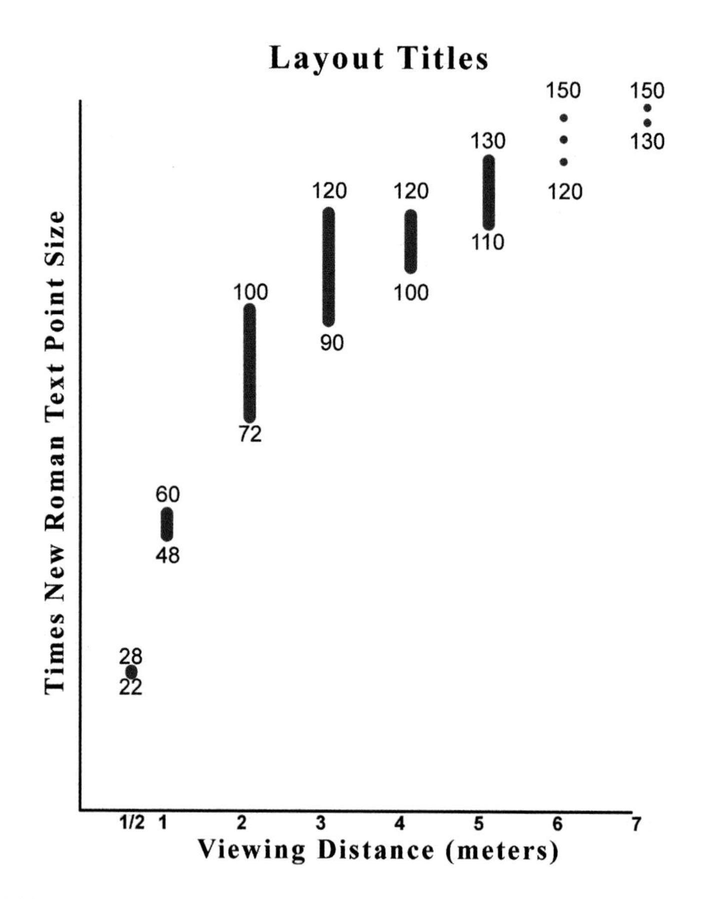

FIGURE 4.12
Approximate point sizes to use for map titles depending on the expected viewing distance. A map intended to be read at a distance of 3 meters, then, should have a title in the range of 90 point to 120 point if using the Times New Roman font.

for title text, and somewhere between those two for labels and captions. Metadata-type text elements, such as network paths, disclaimers, and other text that doesn't need to be seen by the general audience, can be at a close-range text size such as 12 or 14 point.

Other Modifications

Now that you have a good idea of how your chosen font and font size will impact the readability of your maps and layouts, it's time to focus on the other ways you can modify the text. Whether you are designing a layout, creating labels on a map, or modifying a default legend, there are options for changing the font so that it becomes an intentional part of the design. To start with, there are many ways to provide your text with emphasis. One easy way to achieve it, which is especially useful for titles and feature labels, is the

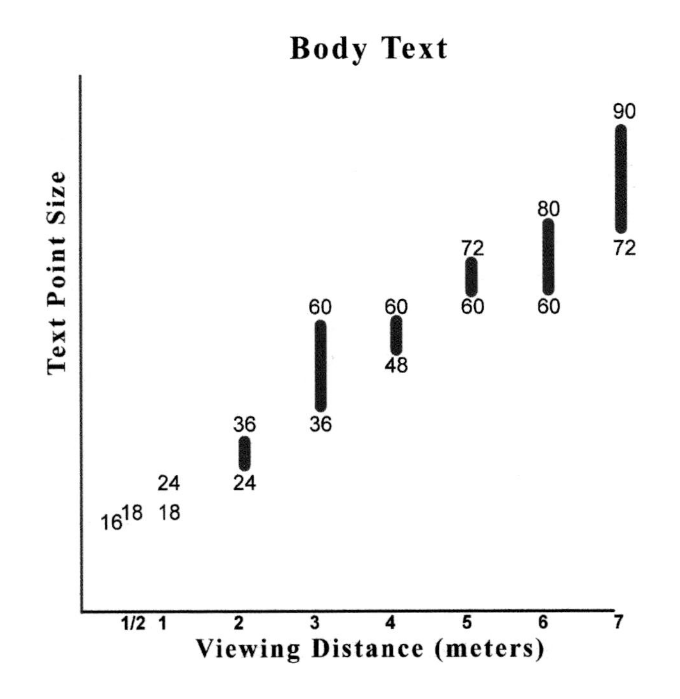

FIGURE 4.13
Approximate point sizes to use for body text depending on the expected viewing distance. These measurements apply to most fonts equally.

use of expanded spacing between letters. This is variously called *character spacing*, *letter spacing*, or *tracking*. Character spacing achieves somewhat the same effect as using a wider font (like Courier), but you can use any font and simply increase the white space between letters (see Figures 4.14 and 4.15).

Italicized text is yet another way to differentiate labels, titles, and other text. The italic font variant was originally developed as a means of creating a slightly smaller sized letter. Italics were intended to be a more compact way of writing without changing fonts. For map labels, italics are reserved for certain features such as streams and oceans. They can help differentiate a title from a subtitle and can also serve as a de-emphasizing mechanism for margin text (see Figures 4.16 and 4.17). Conversely, they can provide *emphasis* on one word in a text block. When an italic font is called for, use a true italic font, not an italicized version of a regular font. Most typeface families contain one or more fonts with real italics, but not always. If the typeface family that you are using doesn't have an italic font, then some software will create a computerized italic version, which is undesirable.

Bold text is commonly used for titles and certain map labels such as city names or other major geographic features. When there are multiple levels of importance for features on the map, the bold text can further emphasize the most important features. One common example of this practice is to label all major cities in a bold font while minor cities are in regular font and towns

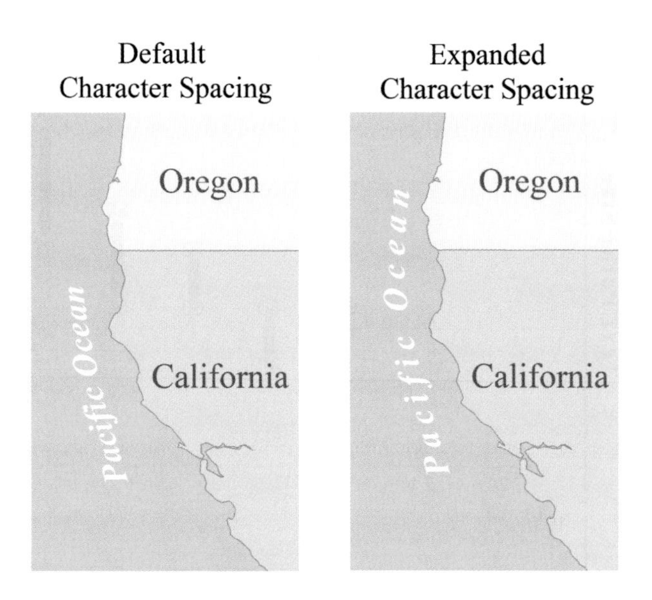

FIGURE 4.14
The expanded character spacing for the text on the right-hand map allows the Pacific Ocean label to be in proportion with the size of the ocean.

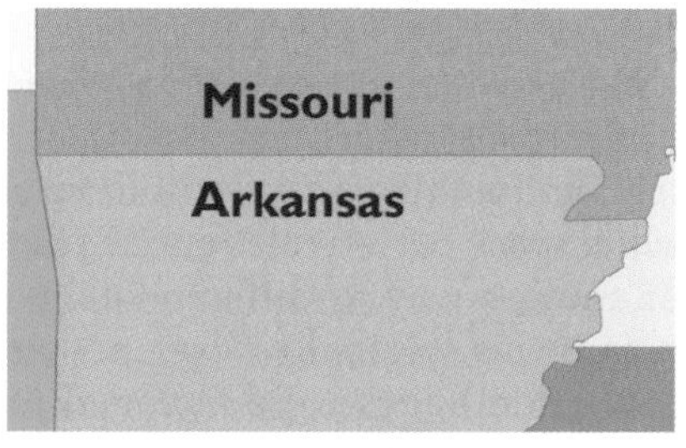

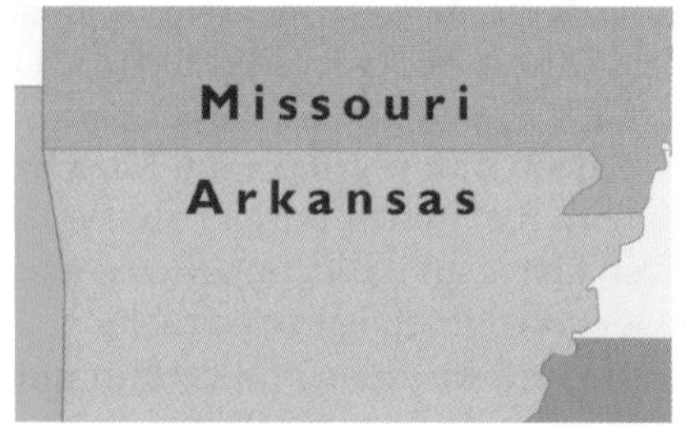

FIGURE 4.15
The expanded character spacing on the bottom map serves to emphasize the state labels more than in the top map, while also being easier on the eye by spreading the bold text out.

The Many Places of Roy Rogers
Cities and Towns in Roy Roger's Westerns, Scaled by Number of Movie Mentions

FIGURE 4.16
The use of italics in the subtitle separates it from the main title and signifies its secondary status to the main title.

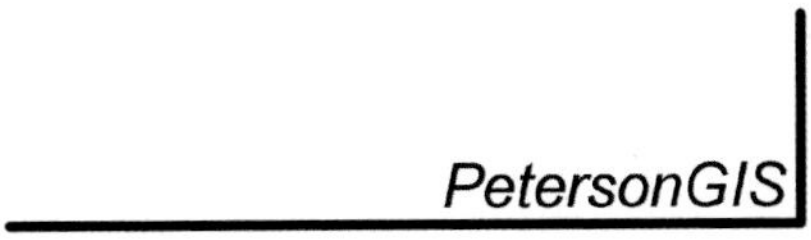

PetersonGIS

FIGURE 4.17
Margin text on a layout that is not central to the map's purpose can be de-emphasized by italicizing the letters.

in a lesser point size. A bold font does *not* increase legibility for small font sizes. A 22-point bold font will be just as unreadable at 4 meters as a 22-point regular font.

Underlining of text is seen much less in GIS, rarely on map labels, and is uncommon on other layout text except perhaps the title. A title and subtitle pair can be underlined, although we usually see other techniques for titles rather than just a simple underline, such as shadow boxes or white space.

In normal body text situations, such as in a disclaimer or a listing of data sources, you may find yourself using a word that is composed of more than one capital letter such as an acronym or the word *DISCLAIMER*. These words wind up demanding too much emphasis due to their capital letters. The easy fix is to slightly decrease their font size. For body text that is in a 16-point font size, try changing the uppercase word to 14-point size. Smaller text would require a commensurately smaller point-size differential (e.g., try a 1-point reduction if the regular text is only 10 or 12 point). Some examples with the main text in 16 point and the uppercase word in 14 point are shown in Figure 4.18.

STANDARDS

There are several conventions that are generally adhered to when it comes to map element text labels. As with the other mapping conventions discussed in this book, if you decide to veer away from them, you at least ought to do so deliberately and with justification! So take a moment to read up on what they are before making such a decision.

Hydrographic feature labels, like the names of rivers, streams, oceans, and lakes, require the use of a script font or a regular, but italicized, font. Initial capitals are always used except sometimes for ocean names, which can be completely in uppercase. These feature labels are often shown in the same or darker blue hue as the feature itself, or in the case of a polygonal water feature like a lake, the font hue is sometimes white. Use white only if it contrasts sufficiently with the blue hue behind it. In many cases that will mean the use of a bold font and a dark blue feature color to accomplish legibility.

The color red is generally reserved for feature labels that are relatively bad, poor, or very important compared to other feature labels. Conversely, green is generally used for feature labels that connote a feeling of goodness, natural, or not important. Browns and greens are normally reserved for area labels like national parks, forestlands, and mountain ranges, though other colors like gray or black are often seen for those features as well. Additionally, elevation labels such as spot heights and contour labels are typically brown.

Mountain ranges are labeled with nonitalic and nonscript font types, are labeled in uppercase, and employ expanded character spacing so that the label runs along the entirety of the range. Town and city names can be in uppercase or initial capitals, with the uppercase lettering usually reserved for very important cities, those with the largest populations, or both. They are not italicized.

In terms of spacing, the distance between a label and its symbol on either a legend or the map itself should be about half the size of the lettering (Figure 4.19).

Also, if you are labeling a large area (such as *Pacific Ocean*) take care to limit your character spacing to no more than four times the letter height so that the letters do not wind up looking unrelated to one another. Another handy rule to follow is that serif fonts are easier to read on text with expanded space. The serifs tend to draw the eye toward the next letter.

DISCLAIMER: Joe County does not . . .

versus

DISCLAIMER: Joe County does not . . .

United States Geological Survey (USGS) data were downloaded in 2008

versus

United States Geological Survey (USGS) data were downloaded in 2008

FIGURE 4.18

In these two examples, the uppercase word is decreased in point size by a small amount to de-emphasize the word. This is useful when the capitalized word is not supposed to be emphasized.

FIGURE 4.19
In this example, the label is in 20-point font and the space between the label and its symbol is exactly 10 point, or half the label size. This rule of thumb applies to labels on the map as well as legend items.

Placing the Text

Correctly placing labels next to or on your features in the map element is difficult although somewhat helped by automation within the GIS.[6] Although great strides have been made in auto text-placement tools recently, it will still be a part of our job for years to come to ensure that the auto placement has indeed placed text where we expected, not deleted text that we didn't want to delete, not overlapped labels, and not cut labels off. Unfortunately, many hours can be spent manually moving each individual label around in order to avoid such things as overlapping with other labels or important features, labels that get cut off at the edge of the page, and labels that are extraneous. Some automated tools are now in existence to help lessen this task, though setting them up properly can take a lot of time, so they are really only useful for large amounts of labeling.[7]

Thankfully, the work of placing text on a layout such as titles, text boxes, and disclaimers, is much easier. The simple rules to follow concerning that kind of text are as follows:

- Ensure that all text is aligned with the surrounding elements including other text, maps, and lines.
- Do not use justified text in poster layouts. Justified text aligns the left and right parts of the text block to the left and right margins and thereby creates a lot of extra space between words. Use left-justified text instead. Left-justified text is aligned with the left margin but jagged at the right margin, keeping the spacing between words constant. Justified text is just too difficult to read quickly.
- Double-space body text on larger layouts to prevent "conference squint" (see shaded box). Having more space between lines of text makes it easier to read.
- Group text elements together. Use the landscaper's Law of Three (three of any plant looks better than two or four): do not separate text out all over the page, but also do not group it all into one single text area. Odd numbers of items in a group look better than

THE THREE LEVELS OF CONFERENCE SQUINT

Level 1: The text on this poster is so close together that the words are blurring together (or is that due to the free drinks?).
Level 2: I can't follow the text; it makes no sense to me.
Level 3: I am squinting so it looks like I am intelligently analyzing this poster when really I'm here for the free drinks.

As a conference poster designer, it is within your power to prevent Level 1 from occurring, it might be possible to prevent Level 2 from occurring, but you do not have much power over preventing Level 3 from occurring.

even numbers of items in a group. For example, create three short paragraphs of text that are each separated by a headline or an extra amount of white space and place these paragraphs together on the right-hand side of the page.

Text Direction

When it comes to placing labels or any other text element on the map, there will be times when the most obvious text direction—horizontal—isn't feasible. Perhaps you need to place a margin element on the layout like a network path (e.g., \cartography\conference_poster\) and you don't have any space for it unless you squeeze it vertically along the side of the page. Or perhaps you are labeling a stream that flows upward in your map. For these types of cases, use the diagram in Figure 4.20 as a guide to help determine how the text can be placed in the most readable manner. The preferred directions in terms of readability are shown in darker text while the less readable (and therefore less desirable) text directions are shown in progressively lighter text.

Another text direction consideration is the following rule of thumb: try to align text that falls on the outskirts of the map element so that it faces inward toward the center of the map instead of outward. However, there are always going to be situations that provide a challenge for the mapmaker. For example, let's say you have a river that runs north–south in the upper right-hand side of the map. You then have a conflict whereby the rules of readability conflict with the rule that says text should face inward. The best thing to do in a situation like this, again, is to be aware of the rules and make sure you at least try both options. Choose the direction that looks the best in the context of your unique data and layout.

FIGURE 4.20
This graphic is a tool to help you determine which way a label or other type of text ought to read on the printed page. The bolder the text for the word *Mapping*, the better the placement in terms of readability.

Text placement and direction rules are illustrated in Figure 4.21. Stream labeling above the line is better than below the line, and toward the center of the page is best. Label halos are used when a label must cross another feature to separate it from that feature. Halos should be kept as small as possible and the color should match, as closely as possible, the main background color (more discussion in the next paragraph). Water features such as rivers, canals, and bays are usually labeled in italics or sometimes script. The title for a static map should be in an easy-to-find location, with the upper left-hand spot being the best for a map of this scale that doesn't have margin elements.

The general consensus among designers is that halos around text detract considerably from a design. The most common reason that mapmakers employ the halo technique is to provide added emphasis on text or to make the text legible on highly saturated and complex backgrounds.

However, when a halo—particularly of the "large and lumpy" type—is used, it subverts and obfuscates the typeface designer's original work. No longer do the spacing, shape, consistency, height, vertical stress, apex form, incline, and counter size that were built into the font make sense. In short, halos typically look bad. As with any "rule," this idea of halos looking bad is more nuanced than this. Small halos, where the color of the halo matches the map's background color, *can* be effective in allowing the text to be more separated from other intervening layers such as roads, trees, and elevation.

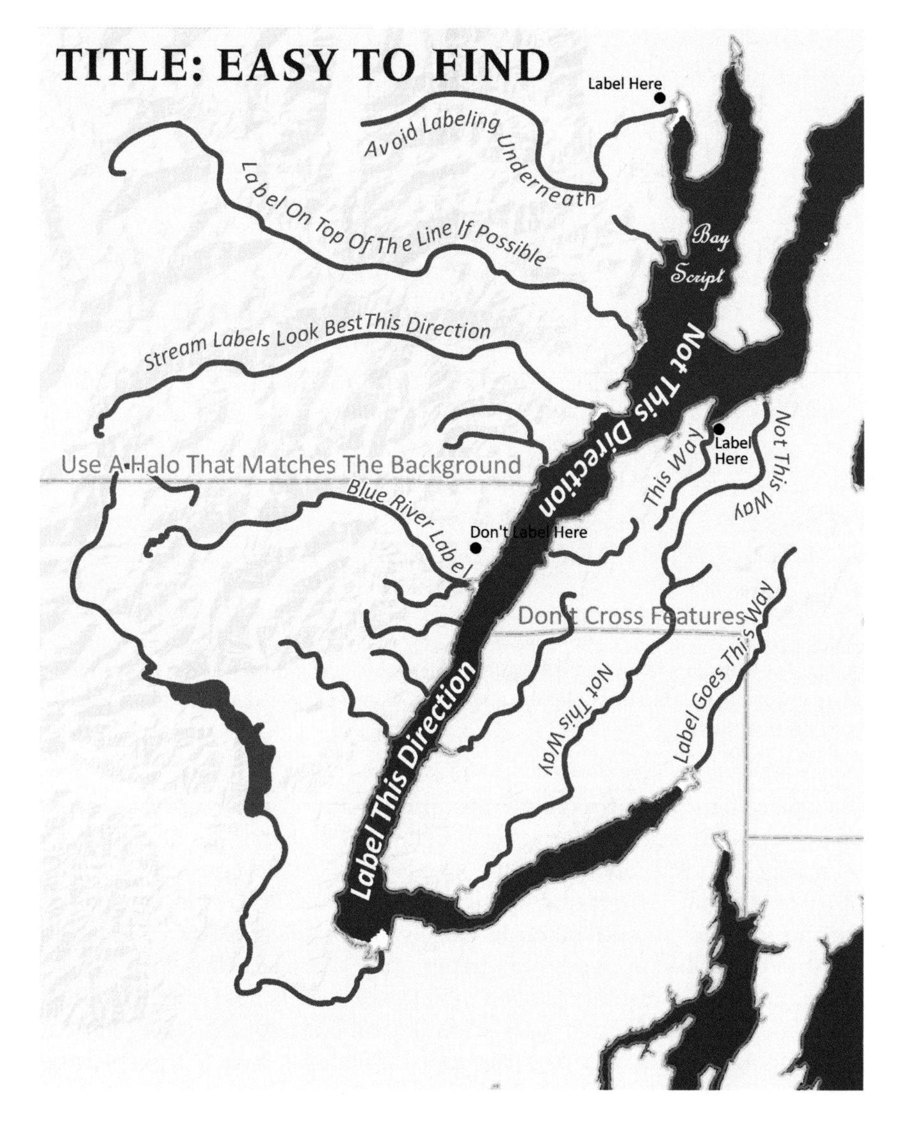

FIGURE 4.21
This map illustrates some common labeling standards in terms of direction, placement, and color.

Let's see this in action. Just to make things interesting, the examples shown here showcase a difficult design conundrum: it involves the visualization of a complex tree height LiDAR analysis, a riparian corridor, and a creek name label. In more simple cases than the one shown in these examples, first consider making the font bigger or changing its color to achieve the kind of emphasis and contrast that you need, before considering a halo. If a halo is still warranted, use the technique shown here (see Figures 4.22 through 4.26).

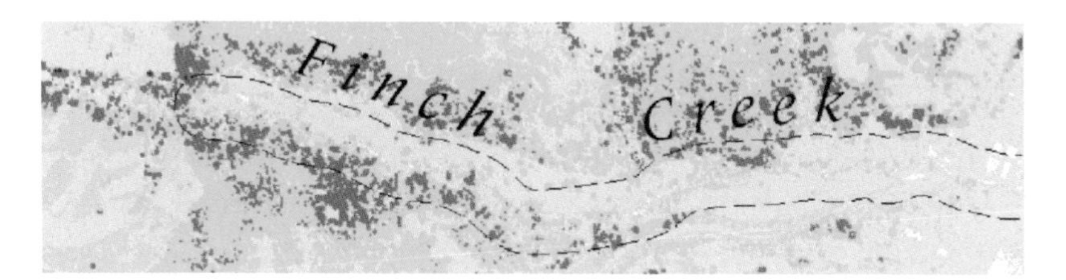

FIGURE 4.22
The initial setup with no halo. It is difficult to read.

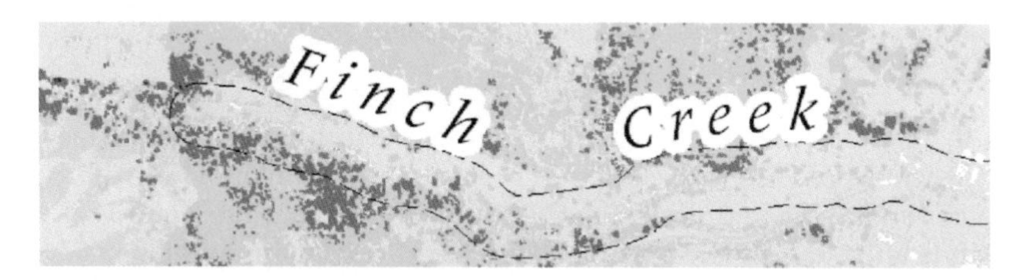

FIGURE 4.23
To make the creek name stand out more, you might try a mega halo. This results in visual catastrophe.

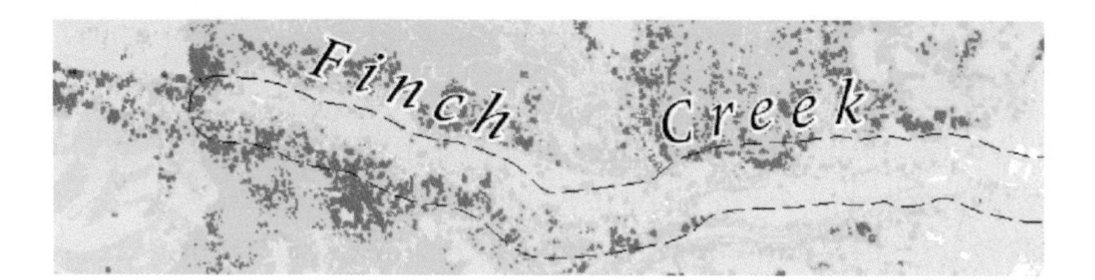

FIGURE 4.24
Another option is to use a smaller halo. Most halos default to white, as in this figure. This still doesn't work.

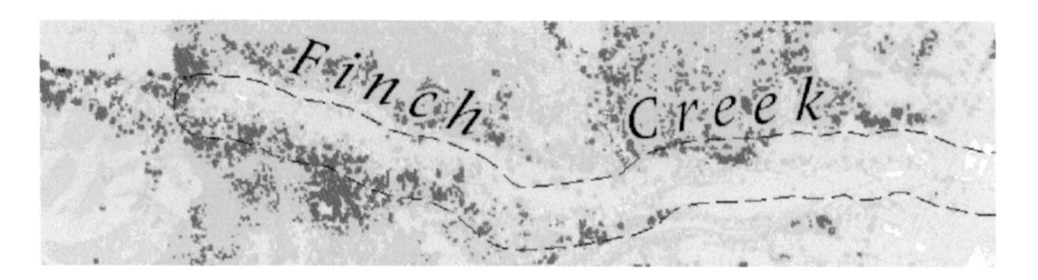

FIGURE 4.25
Using a color for the halo that matches the underlying color—in this case green—is very effective in separating the text without causing undue emphasis.

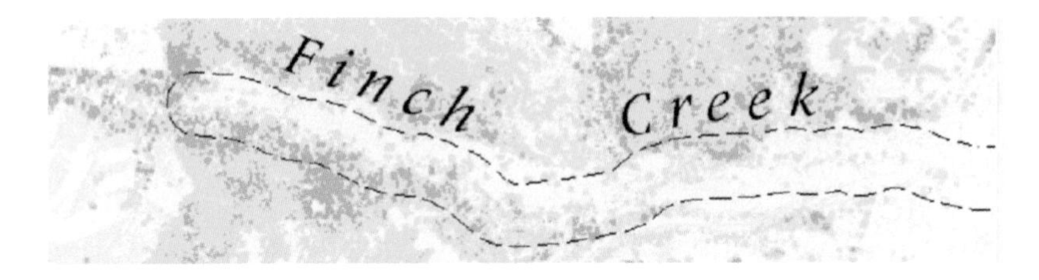

FIGURE 4.26
If using a color for the halo that matches the underlying color—as in the Figure 4.25—will cause a problem in interpretation of the map, then another option is to simply lighten the background enough to allow the figure–ground contrast to be heightened. The cartographer must choose the option that causes the fewest interpretive issues.

With this knowledge of fonts, font-emphasis techniques, point size, and text direction, you are well equipped to produce a fine map where the words both communicate and look superb. You will find that the quality of your text style will enable your map viewers to read your maps and layouts better than they would if you simply stuck with the GIS defaults. In fact, according to some recent research, aesthetically pleasing things actually work better than unattractive things.[8] Keep that in mind the next time you design!

Study Questions

1. What is the difference between a sans-serif font and a serif font?
2. What might a script font be used for?
3. Choose three prominent map features in your locality. Decide how they should be labeled using the label rules from this chapter. Choose a color, character spacing, and font (italic, serif, sans serif) appropriate for each.
4. What type of font is useful for vertically aligned numbers, and what type of font is useful for numbers that are in-line with text?
5. How wide can character spacing get before text becomes too difficult to read?
6. List three river labeling cartographic standards.
7. What color should a text halo be if possible?
8. What point size is roughly equal to a one inch letter height?
9. Describe some labeling standards for city and town labels.
10. What type of text is most legible at a distance?

Resources

Dafont.com is another user community site that allows free downloads of unique fonts: http://www.dafont.com.

Font List allows you to see font samples from a clickable list of fonts: http://www.fonts.com/fontlist/.

FontStruct is a user community site that allows you to build your own font and download other users' fonts: http://www.fontstruct.com.

Linotype is a site that contains original, proprietary fonts for purchase: http://www.linotype.com.

WhatTheFont identifies a font from a user-uploaded screenshot of the desired font: http://www.myfonts.com/WhatTheFont/.

Endnotes

1. There is a tradition of naming a font (typeface) after daughters, and it is also not uncommon to name a daughter after a typeface. Thank you to Matthew Carter for confirming this detail (Matthew Carter, Re: Georgia Name, E-mail to Gretchen Peterson, June 29, 2008).
2. A. Z. Zineddin, P. M. Garvey, R. A. Carlson, and M. T. Pietrucha, "Effects of Practice on Font Legibility," *Perception and Performance 4, Human Factors and Ergonomics Society Annual Meeting Proceedings* (2007): 1717–1720.
3. Wild West Shadows font, 2001, West Wind Fonts, http://moorstation.org/typoasis/designers/westwind/ (accessed October 8, 2013); Dieter Stefman, Olde English font, 2006, http://www.dafont.com/olde-english.font (accessed October 8, 2013). The Outer Space font is not available online.
4. Freestyle Script font, comes with Windows. Manfred Albracht, CommScript font, http://desktoppub.about.com/library/fonts/hs/uc_commscript.htm (accessed October 8, 2013). Exmouth font, Prima Font, http://desktoppub.about.com/library/fonts/hs/uc_exmouth.htm (accessed October 8, 2013).
5. Special thanks to my husband for being a second pair of eyes and confirming the results.
6. H. Freeman, "Automated Cartographic Text Placement," *Pattern Recognition Letters* 26, no. 3 (2005): 287–297.
7. Dymo is one such product for better automated label placement, created by Michal Migurski.
8. D. A. Norman, *Emotional Design*. Cambridge, MA: Basic Books, 2004, 17–20.

5

Color

The color on a map layout is like frosting on a cake. Sure, the frosting is the best part of a cake, but a lump of pure frosting without a nice, well-made cake under it has no reason for being. The same goes for a map layout decorated with great colors but with no clear underlying message. That's why good color choices are nice to have, but aren't the most important aspect of your maps and layouts. By telling you that, my main motive is to take away the fear that sometimes gets hold of us analytical types when we have to do something artsy. If color can be proved to be of secondary importance, then perhaps it will not be so intimidating. Of course, this doesn't mean you can ignore color altogether. No, we need both color and the underlying substance. To continue with the analogy, a cake without frosting is no fun at all.

Let's start the argument about color not being *that* important with this question: Is color what you remember about the most famous maps you've seen? No? What is it about those maps that you remember most, then? Was it the underlying message, the really well-organized features, the immaculate presentation of the information, the dragons in the ocean? Some of the most ubiquitous and well-regarded maps in the United States are the US Geological Survey's topographic maps. Yes, I can recall off the top of my head that those maps have brown contour lines and possibly a green background for certain features, but I don't recall any of the other colors on those maps. I also probably couldn't tell you what colors are on Google Map's interface, but I use it all the time.

So color provides supporting information to well-made maps and layouts. Certainly, if those Google Maps were made with garish colors they would not be as popular as they are. The purpose of color on a map is to help a map viewer decipher the symbols in order to make meaningful inferences. Color is not, by itself, the purpose of a geographic information system (GIS) map. Even when color is used to display a variable, it is still only offering a means toward the visualization of that variable. For example, a map of watersheds could show some watersheds in green and some in black depending on the amount of intact forestland in each. Even though color is important in a map like this, as it serves more than a purely decorative purpose, the meaning of the map is still centered around how much forest cover is in each watershed, not around the nice colors.

Because color is supporting information, we can conclude that poor color choice does not always affect the map in terrible ways. Those of us who haven't an ounce of color theory in our background or a firm knowledge of color connotations may still produce a map that somewhat conveys its point, albeit perhaps in an ugly way. So we may produce a map that effectively communicates but is not necessarily aesthetic. However, the chances that the color palette doesn't impede the map reader's understanding are slim and it's not a chance you want to take. A professional, compelling, and communicative map nails the analytical component while also standing on its artistic legs.

What are some ways that poor color choice can impede understanding? One way is explained in the "Five Shades Are Enough Already" section of this chapter, which deals with the problem of maps that have too many shades of the same color. This can leave a map viewer completely unable to match the colors used in the map to the colors in the legend. We've already mentioned that a jarring color palette can lead a map user to abandon using it. A color palette that seems dated can also discredit the map, if it seems as if it is not current enough. In short, there are many ways that color can become an impediment if it's not done right.

One caution: creating an aesthetically positive map is an art, and as such, it is inherently subjective. You will see this exemplified time and again in your career. Let's say you've just spent two hours picking out colors via online tools, inspiration pieces, looking through your GIS color defaults, and so on. You then take your extremely well-thought-out map to your colleagues or bosses and two-thirds of their feedback focuses on how they wouldn't have used pink for such and such feature or how your red isn't quite the right shade. Everyone is a critic when it comes to the subjective art of color picking! That's because it is one of the easiest things for a nongeoprofessional to focus on when asked to critique a map. While you may never be able to completely satisfy everyone with your color schemes, you might, through some time spent learning about the basics of color, and a lot of time spent applying these concepts to your map, and a smidgen of luck, be able to get the color critiques down to about one-third of the overall discussion. If that can be achieved, then you have succeeded.

HOW TO PRESENT YOUR DRAFT MAP TO THE BOSS OR CLIENT

When you have to submit a map for review, feedback, or final approval, you need to be confident and able to explain the exact reasons why you made the map the way you did. First of all, you will want to explain any modern styling that you used because anything that is new and different from the old approach will cause initial uneasiness. People who aren't involved in map making on a daily basis will assume that map styles from 10 years ago still work today. They don't realize that the art of making maps is constantly evolving as technology changes and color trends change.

Therefore, explain your reasoning up front. If, for example, you created a dot map instead of the gradient-color scheme they are used to, you can explain that dot maps are actually a better representation of the features for your particular map and why. Another example would be if you used a nontypical color for some text (like pink). You would want to point out that you tried many colors and it turned out that pink was the only one that would create enough contrast to make the text visible from a distance, or whatever your particular reasoning was.

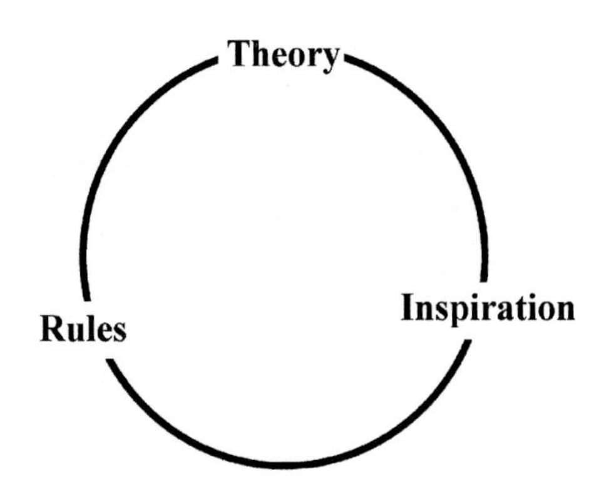

FIGURE 5.1
The Circle of Color Love illustrates the three overarching color concepts. You might wonder why it is fashioned into a circle instead of, say, a triangle. The answer is because a love triangle is inherently discordant while a circle is a symbol of unity, and indeed when all three of these items are properly considered together, they will bless you with a *love*ly map!

Now, while you are definitely on your own when trying to respond to those color comments, this section can help with your first two goals: learning about color theory and figuring out how to apply it to your GIS map. This chapter is organized into the three aspects of map color that are the most important: theory, rules, and inspiration. There is no real hierarchy among the three concepts. That is, they are all equally important. These concepts are illustrated in the Circle of Color Love diagram in Figure 5.1.

Color Theory

Color theory centers first and foremost on the color wheel invented by Sir Isaac Newton way back in 1666. You surely have seen the classic color wheel, or one of its derivatives, before. Whichever color wheel you have used or seen probably shows the 3 primary colors, the 3 secondary colors, and the 6 tertiary colors, with 12 in all. The static color wheel in Figure 5.2 illustrates this. There are many interactive color wheels on the web that are very useful when it comes time to put together a map palette. See the "Resources" section at the end of this chapter for references to several interactive color wheel options.

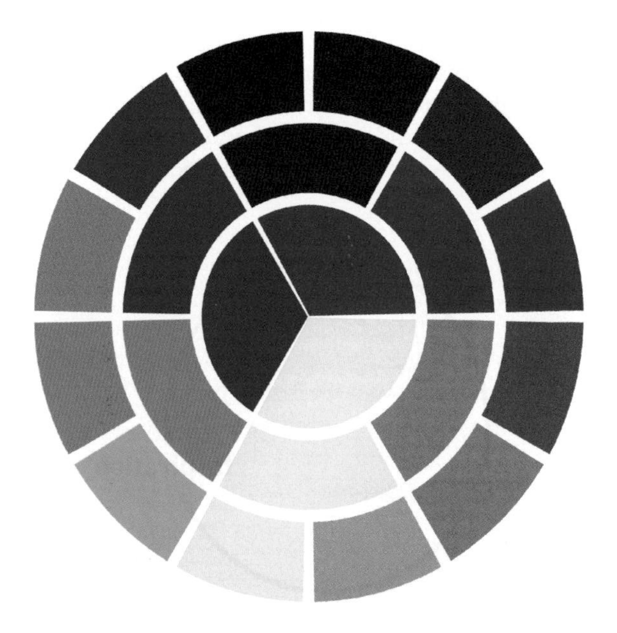

FIGURE 5.2
This color wheel shows the 3 primary colors in the center (yellow, blue, red), then adds on the 3 tertiary colors in the middle, and then adds on the 6 tertiary colors in the outside circle.

You can apply the color wheel to the work of picking out colors for a map in several ways, referred to here as *color harmonies*. The color harmonies are combinations of colors based on their positions in the color wheel. They are *analogous*, *complementary*, *polychrome*, and *neutral*.

Analogous color palettes are similar to one another, and are close to each other on the wheel. A palette created with these types of colors will create a subdued, quiet effect on the design. Color schemes with light green and blue, for example, are ubiquitous in reference map design and, if done right, are calming to the eye.

Complementary color palettes are opposite to each other on the wheel. The traditional complementary color pairs are red and green, blue and orange, and yellow and violet. These colors will create a vivid color scheme that demands attention, but is at risk of being overly ostentatious. Many fast food signs contain complementary colors to attract attention, for example. In mapping, complementary colors are often used in choropleth and heat maps. These are sometimes referred to as *diverging* color schemes when they are also accompanied by a hue progression from light to dark or from dark to light. In reference mapping, complementary colors are employed in order to provide sufficient contrast. For example, a map could have a green background and red reference lines.

Polychrome color palettes include many colors from all over the color wheel. These need to be carefully chosen via trial and error, past experience, or an inspiration piece (more on this last concept later in the chapter) in order to ensure that they don't clash. Maps with polychrome schemes are common on layouts where multiple layers are visible at the same time, as the use of many colors is sometimes the only way to differentiate the layers adequately. This schema is also used on maps where a single layer needs multiple feature-type differentiation. These schemes need to be visually cohesive despite their many and varied hues. Mapmakers may use color schemes that are already proven, by using reference materials such as paintings, color palette books, and the like, to find a good palette. Applying the stronger colors to smaller features is one method of achieving color harmony in such a scheme. The reverse—strong colors throughout with a few, smaller, bright colors—is also sometimes effective.

A *neutral* color scheme contains only black, gray, and white. Sometimes this definition is expanded to include colors that are particularly unsaturated, or absent of hue, such as beige and taupe. In cartography, neutral color schemes are used in report maps or other publications that will be printed on black-and-white printers. However, they are not unheard of in web map design. Neutral color schemes are discussed more in the "Grayscale Modern" section at the end of this chapter.

Digital, interactive maps use a hexadecimal (HEX) system for defining colors, while paper maps use any of several other color definition systems. Most GIS software will give you multiple options for defining any custom colors you use. Some of the options that your GIS (or graphic design software) may give you are red, green, blue (RGB); hexadecimal; hue, saturation, value (HSV); cyan, magenta, yellow, black (CMYK); and CIELAB. The following paragraphs present brief descriptions of each of these models so you can decide which would be best suited for your projects.

RGB

The RGB system stands for red, green, blue and was created for electronics like cathode ray tube computer screens and color TVs, where a monitor emits differing amounts of red, green, and blue to produce a desired color. Think of RGB as a means of describing a particular color in terms of the amount of each primary color you would need to put onto a black surface in order to produce the color. Because the initial surface is black and you are adding primary colors to it, the system is described as an *additive* color system. The color is managed by specifying a number between zero and 255 for each of the three primary colors. So if you were to put a full amount of red onto the black surface but no green or blue you would wind up with an RGB of 255 0 0 (red). As the number decreases, the brightness of the color decreases. So an RGB of 150 0 0 will still be red, but a darker red than 255 0 0. These numbers are referred to as *RGB triplets* throughout the book. More RGB triplet examples are shown in Figure 5.3.

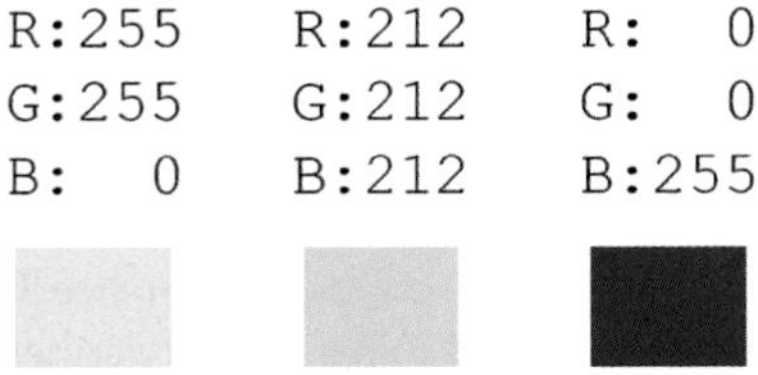

R:255	R:212	R: 0
G:255	G:212	G: 0
B: 0	B:212	B:255

FIGURE 5.3
RGB triplets are shown here for yellow, light gray, and blue.

With most GIS software, the map export function produces files with colors specified in RGB. This is important to know about your software if the quality of print output is important to you because most printers will need to convert the RGB to a different, but related, system (CMYK) prior to printing. That conversion process can change some of the colors that don't convert well from one system to the other. Remember: the RGB system is for screens, while CMYK is for printers. More discussion of CMYK appears after the hexadecimal and HSV discussions in this chapter.

Hexadecimal

The hexadecimal (hex) system is a general term describing a numerical system with base 16. In it, the numbers 0 to 9 are paired with the letters A through F to provide 16 total characters. Hex is used for many purposes other than color coding, but if your work involves creating digital, interactive maps, you will need a good understanding of the hex color system, which is a derivative of the RGB color model. Whereas the RGB colors are specified by means of three numbers, the hex system uses three *pairs* of letters and numbers. Hex is specifically used for hard coding colors for display on a device.

In HTML, the hex letter-number pairs are preceded by a "#" sign, and written like: #83F52C (greenish). There are 256 potential brightness values for red, 256 for green, and 256 for blue. So, if you want to specify a brightness value when you are displaying color on a monitor, you need three brightness numbers paired with three color numbers. The hex system allows this. The zeros in hex represent a no-color value while the Fs can be thought of as full-color values. So if you want pure green, the hex notation is #00FF00, which means that the entire green component is displayed while none of the other components are displayed. The brightest colors are represented in pairs of Fs: FF, while the darkest colors (or absence of colors, since this is an additive system) are represented with pairs of zeros: 00. The whole range goes: 00, 33, 66, 99, CC, FF. See Figure 5.4 for more hex examples. These are just the basics. To read more on the very exciting world of hex colors, please see the "Resources" section.

FIGURE 5.4
These are examples of the colors light green, maroon, and dark gray, as written in hex notation.

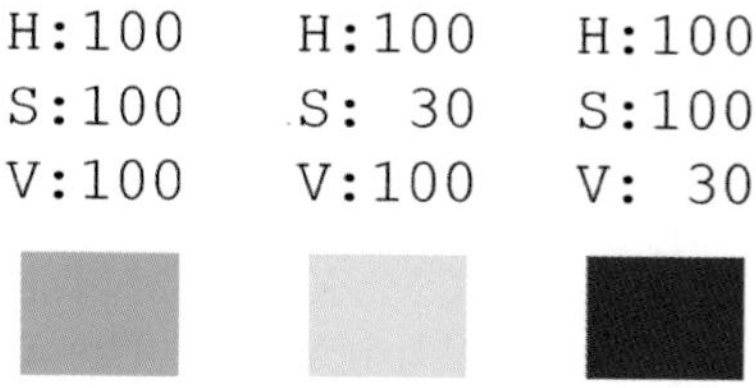

```
H:100   H:100   H:100
S:100   S: 30   S:100
V:100   V:100   V: 30
```

FIGURE 5.5
These shades of green are represented by the HSV color system. The hue is 100, denoting a green color, the saturation varies from 30 to 100 percent with more saturation being a stronger green, and the value varies from 30 to 100 percent with more value being a lighter green and less value bringing the green closer to pure black.

HSV

The hue, saturation, value (HSV) color system is similar to RGB in that you specify a set of three numbers for each color. Each number represents a principal component: hue, saturation, or value. The hue (color) is a number between one and 360, the saturation (amount of gray in the hue) is a number between one and 100, and the value (amount of white in the hue) is a number between one and 100. For the saturation, the number represents a percentage of pure color where 100 is 100% pure color and 0% is gray. For the value, the number represents a percentage of brightness where 100 is 100% white and zero is 0% black (Figure 5.5 shows a few examples). This is a handy system to use when you want to keep the basic hue while tweaking the saturation and value.

HSL

The hue, saturation, lightness (HSL) color model is the same as HSV in the hue component, but differs in the last two components. The saturation part is defined differently than in HSV, and of course the last component is entirely different, with the *L* standing for *lightness*. In either case, these are the color models you want to use if the brightness and saturation of a color are important to you. For example, if you want to change the brightness or saturation in RGB or CMYK, you would have to change the numbers for *all* the values, which can also cause the hue to change, whereas with HSV you need to adjust just one number for either the saturation or the value, which keeps the hue constant. These models are also slightly more useful than RGB in that they actually contain a more complete set of colors that are visible to humans than RGB, which is slightly restricted.

CMYK

The cyan, magenta, yellow, black (CMYK) model is sometimes presented as another option in your GIS for defining colors. The *K* stands for *key*, a term that is a relic from the printing press days when the three colors would be aligned or "keyed" with the black ink in order to save on ink costs. The K is not always included in the acronym. You can think of this model as being a white sheet of paper (which reflects light back at you) that becomes a color once you add color pigments to it. Because each pigment you add to the white paper is actually taking away wavelengths of the reflected light to produce the desired color, it is called *subtractive*.

The numbers associated with each of the principal components of CMYK are in percentage format. That is, if you specified 100% for C, M, Y, and K, you would wind up with black; and if you specified 0% for C, M, Y, and K, you would have white. Some other examples of CMYK combinations are shown in Figure 5.6.

When CMYK is an option in your GIS, you may want to use it if you are concerned about matching the look of the color on the screen with the color on a printer. However, you have to be cognizant of how your GIS exports color. Just because you defined a color in CMYK does not necessarily mean it will be exported in CMYK. In fact, some software will convert your color definitions to RGB when you export. This is mainly a concern only for those people who are sending their file off for offset (commercial) printing. If your software exports in RGB, a way around it is to export to graphics software like Adobe Illustrator or Adobe Acrobat, re-change any colors that changed,

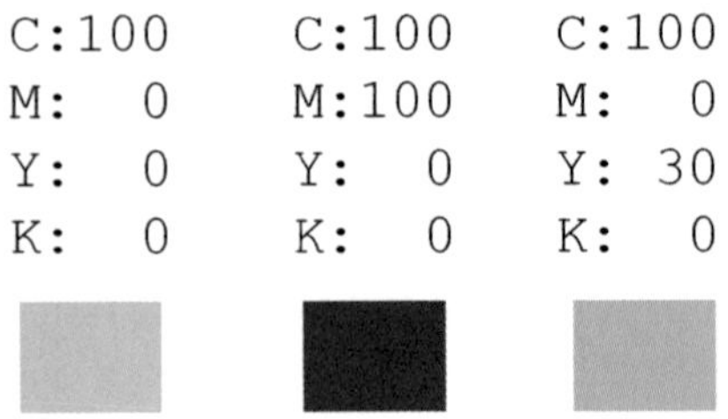

FIGURE 5.6
These are some simple CMYK values and their associated colors.

and then export to a printing file format from there, specifying CMYK mode. Another option is to print out a conversion sheet that pairs what a color looks like on-screen with what it will look like on a printout.

CIELAB

Another color model, variously referred to as CIELAB, Lab, or CIE (L*, a*, b*), is distinguished from RGB and CMYK in that it describes all of the colors that are visible to us, as opposed to a slightly restricted set. It also contains colors that *aren't* visible to us (imaginary colors, if you can believe it). It is a three-dimensional (3D)-based model where the xyz coordinates are defined by the components L, a, and b, where L refers to the amount of lightening applied to the color.

You won't often find the CIELAB color model in the GIS realm, but it is good to have at least a cursory understanding of it. Some examples of the values and the colors they create are shown in Figure 5.7.

Rules

So what do we need to learn about color, other than color wheel combinations and the various color models used in our software? Color rules. It may seem strange to think of rules that can be applied to color since choosing and combining colors is highly subjective. However, there actually are some rules that are particularly important for geoprofessionals to understand in order to ensure the communicative abilities of our maps. The two most important rule sets deal with figure–ground and the viewer's perception. Figure–ground is discussed first. Next is the viewer's perception, which is laid out as a series of rules that includes the five–shade rule, color gradients, color connotations, blending, contrast, quantity, and color deficiency.

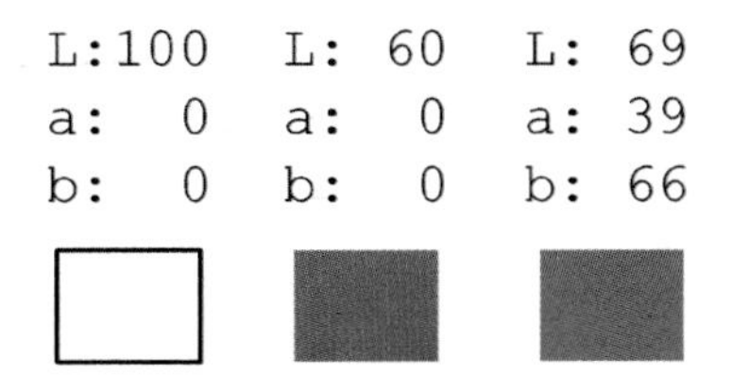

FIGURE 5.7
These CIELAB examples illustrate the lightness value as it goes from high to medium (white to gray), and then how the gray will change to a color as soon as you add values for a and b.

Figure–Ground

In the term *figure–ground*, the word *figure* refers to the foreground, or focus, of the map while the term *ground* refers to the background. For a map of topography of the Galapagos Islands, for example, the islands are the figure and the surrounding water is the ground. If the map is of the bathymetry surrounding the Galapagos Islands however, the focus would be switched and the water would be the figure and the islands would be the ground.

Gestalt is a general term that describes a group of objects (physical, biological, or even psychological phenomena) that have a definition as a group that is different from their definitions when they are apart. This concept, borne of German psychologists in the early 1900s, when applied to graphic design, encompasses many concepts including image continuity, closure, similarity, and figure–ground. For the purposes of this discussion, we are interested primarily in the gestalt concept of figure–ground, of course, which refers to the differentiation between an object and its background. GIS maps usually include objects that need to be emphasized and separated from the other objects on the map, even though the other objects are also important for geographic context. This applies to feature pairings such as land and water, city points and land, or watersheds and forest stands.

There are many ways to emphasize certain objects on a map. One of the easiest is to highlight the borders between the two objects. The classic example of this technique is the application of a dark blue color to the edge where water features meet the land features. The dark blue color would necessarily be a darker blue than the water feature itself in order to create the necessary contrast. Other examples include differentiating the subject from the background via a contrasting color (see Figure 5.8) or the inclusion of vignettes (sometimes called banding) along the shared feature edges (see Figure 5.9). Historic maps often employ the vignette technique. Text labels also benefit

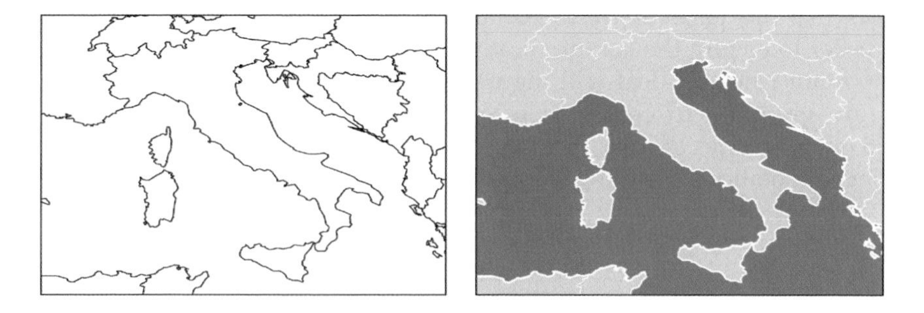

FIGURE 5.8
The map on the left does not show enough contrast between the features for the map viewer to understand the map. The map on the right differentiates water from land with both color and a highlighting mechanism (white color) along the water–land interface. A dark blue is typically used as the highlighting color, but in this example white works well due to the dark hue of the land and water. Some other great figure–ground color combinations are gray and blue, light blue and dark blue, and light green and dark green.

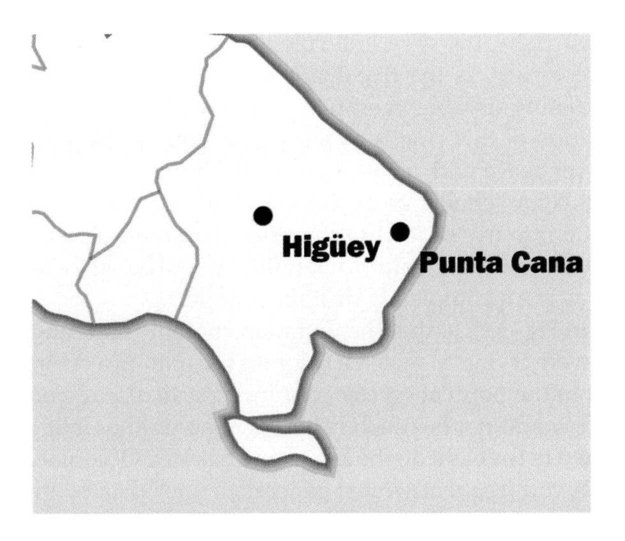

FIGURE 5.9
A vignette, or gradual fade-out, between water and land to emphasize the contrast between water and land.

from figure–ground differentiation. A halo, for example, around black text labels is a common technique for making the text appear to pop out from the background of the map (text halos are also discussed in Chapter 4, "Fonts"). The goal of these techniques is to provide the needed differentiation between objects so that the map viewer can accurately distinguish what is what.

A more extreme method of differentiating objects is applying 3D effects, or extrusions, to single objects on the map. When this is applied to a 2D map, however, it is important to remember that the map is no longer a "true" map since at least part of the data has been arbitrarily removed from coordinate space. For example, an overview map of a study area in southwestern Colorado might show Colorado and the surrounding states, but with the state of Colorado offset from the others in order to emphasize it. It is an effective technique in that example because, even though the extruded state of Colorado obscures the state(s) behind it, it still can provide a reasonable overview of where the study area is. There are many media in which this technique can be effectively employed including reference map graphics, report covers, and icons. Additionally, web map graphics can utilize the technique in order to emphasize certain clickable parts such that when a user hovers over a particular region, the region "pops out" at the user to make it easier to click. Obviously, the use of this technique results in strictly graphics-type maps instead of analytical-type maps, since they can obscure important data.

Five Shades Are Enough Already

If your map viewer cannot interpret your map at all or interprets it incorrectly, then you have failed in your pursuit to communicate. In order to avoid this catastrophe, the very first rule to learn in the viewer's perception category of color rules is the five-shade rule. It goes like this: the human eye can only distinguish between five shades of the same color (hue). The correct application of this rule is as follows: if you are emphasizing change in a variable over space with a commensurate change in saturation of a color (via a gradient color scheme as explained in the next section), you need to group your data into five or fewer categories. You must also take care not to use any shade of that same color to denote any of the other features on the map unless they are a completely different shape.

You can see in Figure 5.10 that the first map contains difficult-to-distinguish shades. If you were trying to match a color on the map to a color on the legend to, say, figure out the population category for a particular region of Spain, the map on the left would not be much help. The map on the right, with only five shades compared to the eight on the left, is much better. Of course, applying this rule means that you frequently must generalize your data by grouping it into fewer categories, resulting in some loss of detail. Therefore, if it is essential that you have more than five categories, you must use a color scheme that includes multiple hues or even other textures, such as shading or hashing, to further distinguish the colors. If you must use hashing, please note that it often creates a garish-looking map. To avoid this, hashing is to be used sparingly, on the features with the least amount of map area. Also note that in the above example you would not use the same color for any other map element features, such as labels or town points, because again, they would be too difficult to differentiate from the background.

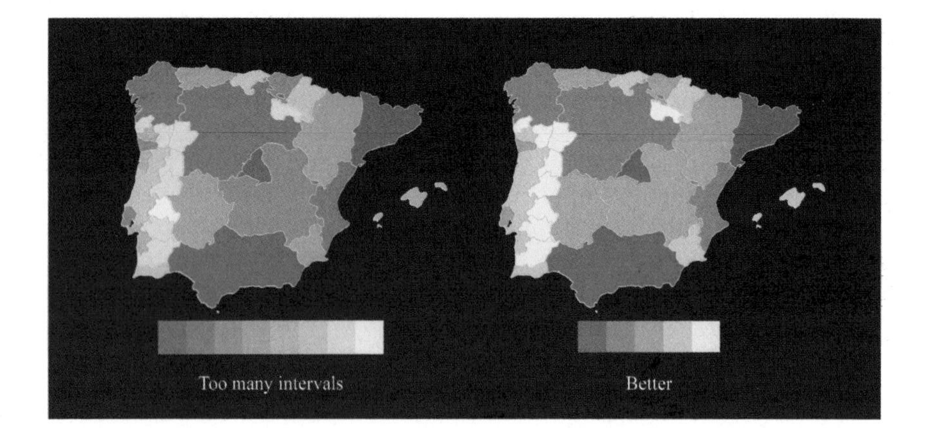

FIGURE 5.10
While we can spot general trends in the map on the left, we cannot determine the value of an individual polygon easily. Moving our eye from the map down to the key and locating the matching shade is virtually impossible. The map on the right shows a generalized version of the same data, but the trends are still visible *and* we can determine the category for each polygon.

However, that's not to say that you can't repeat the same shades in a *layout* containing this map element. Indeed, it would be very nice to bring the same color into some of the surrounding elements, such as the title, border, or layout background, in order to further emphasize your color scheme. This is sometimes called *color echoing*, meaning that you are echoing the color from one part of your map into a part of the layout to provide a cohesive visual look. This discussion about the five-color rule leads us right into a description and best practices of color gradients, and maps that use color as their main form of communication.

Color Gradients

A common type of map that utilizes a color gradient is the *choropleth* map. Everybody has seen a choropleth map before, but many people have never heard the word. Choropleth maps illustrate the value of a variable across the landscape with color that changes across the landscape, within particular geographic areas. For example, a map of crime rates across the country could be shown via a choropleth map by using a yellow to red color scheme, where yellow is used in counties with low crime rates, light red in counties with medium crime rates, and dark red in counties with high crime rates. By showing the variable with these three different colors, we can quickly scan the map to answer such questions as: Is crime higher in coastal areas versus inland areas? Where are the hotspots, or concentrations, of crime? Is my hometown in one of them? Note that the same map data expressed as a continuous crime rate (perhaps via a kernel density procedure), *not* binned into counties, would be correctly categorized as a *heat map*, not a choropleth map. Heat maps also use color as their primary means of communication to the map reader, usually to denote geographic hot spots in data, but represent much more precise hot-spot locations through the use of algorithms applied to nongeographically discreet data.

In both choropleth maps and heat maps, the intensity of color, either through saturation percentage or through the hue, denotes intensity of the variable. The resulting palette is referred to as a *color gradient* or sometimes *color ramp*. One of the things you can do to make your choice of colors easier is to think about the variable you are trying to map and whether there might be a literal or metaphorical color to complement it. For example, is your variable household wealth? Maybe you would want to use a light green–to–dark green color gradient since green is often associated with money. The light green–to–dark green color gradient is a single-hue gradient. A simple example of a choropleth map that uses multiple colors might be a map of college swimming programs organized by state. States with the most college swimming programs might be shown in a dark blue color, states with fewer college swimming programs could be in a lighter blue color, and then perhaps states with no college swimming programs at all would be shown in white or black.

Your choice of palette for color gradients becomes complicated when there are a lot of data values for the variable that is to be mapped. Elevation maps, which are not technically choropleth maps or heat maps, but do employ color gradients, are great examples of how complex the color scheme can get (see Figure 5.11[6]). The change in the variable's intensity (elevation height) is indicated in this example by changes in both hue and saturation. The reasoning for this is that there is such a great range of elevation values in this particular map that more than one color variable is needed (recall from the five-shades rule discussed previously that no more than five shades of a single color should be used). You will note that the colors are also somewhat tied to our expectations of colors at particular elevations. The white at the highest elevations evokes ice caps and the green in lowland areas evokes meadows or lush farmland. This is a common practice for elevation maps and is referred to as *hypsometric tinting*.

Again, thematic maps that indicate a change in a variable within predefined geographic areas such as zip codes, countries, or watersheds are called choropleth maps. One of the traps that can occur with this type of map is that you can easily misrepresent the underlying data. If you want to show how the variable changes across an area, but the variable is grouped into features that themselves have variable extents (think countries as opposed to a regular grid), it is important to normalize the variable with respect to the area it is representing. For example, a map of the deer population in each county across the United States would look completely different if you first divided the deer population by the county areas than if you just displayed the raw data. Dividing the counts by the area produces a map of deer per square mile within each county, which allows much more appropriate cross-county comparisons. Other thematic datasets need to be normalized by population so that the choropleth doesn't simply mimic population density. A map of newspaper readership rates that isn't normalized by population would probably look very similar to a pure population map, with readership rates

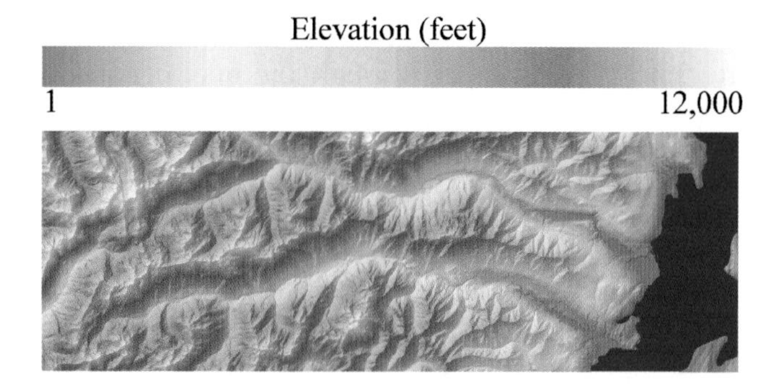

FIGURE 5.11
Because there are so many elevation categories to display on this map, they are shown with both changes in color (hue) and saturation of color. So for low-lying areas we have lighter shades of green that get darker as elevation increases, and for medium-elevation areas we have lighter shades of orange that get darker as elevation increases, and so on.

looking high in the most populated areas and low in the least populated areas. Obviously, if the map is to show something meaningful, normalizing the data by population will then give a better indication of where the readership rates are really the highest and lowest.

Of course, we must make an important distinction between maps that display the variable in its raw state and choropleths that display the variable as normalized by some pertinent factor (commonly area or population, as discussed). You need to carefully consider what your map is trying to show. For example, if you want to show the number of houses by province, use the nonnormalized numbers. But if density of houses by province is really what you are getting at, then normalize the number of households by area. In most cases the appropriate choropleth method is to normalize your data.

DOT DENSITY MAPPING, SYMBOL GRADIENTS, AND OTHER ADVANCED TECHNIQUES

Dot density maps can be a good alternative to choropleth maps because they indicate the density of the raw variable without normalizing it. The map viewer still gets an idea of how the variable is distributed, and in digital, multiscale maps in particular, can perhaps glean more information from them. For example, a population map can show a dot for every 10,000 people in each county, with the dot(s) being placed somewhere inside the county polygons. In this way, the map viewer needs to only look at the differences in the numbers of dots in each county to compare them as opposed to using a legend to help interpret a color gradient. Of course, a legend would still be needed to provide the ratio of dots to people, but it still winds up being less work for the map viewer to glance at that ratio once as opposed to going back and forth between colors on the map and the legend. Recently, dot maps of population have met with critical success by using a one-to-one ratio of dots per person on multiscale digital maps, where the user can zoom in and actually imagine that each dot is one person. See the soils section of Chapter 6, "Features," for another application of dot density mapping.

Another way of displaying variation in feature values is to create a gradient of symbol sizes corresponding to the variable. This works mostly for point layers, but is also sometimes applied to line layers under certain circumstances. A classic point feature example is city points, where the largest cities are symbolized with big squares while small towns are small squares and other towns are in between. A line feature such as roads can be displayed with different thicknesses based on traffic counts. This is different from the dot map method because the size of the symbol is changed, whereas in the dot map method the dots stay the same size and only the number of dots changes. One caution to be aware of when creating a graduated-symbol map is to avoid the use of circles as your symbol if you have more than three or so categories, since small changes in circle size may be difficult for the human eye to detect.[1] The alternative is to use squares, triangles, or some other shape instead.

Some recent maps have employed a hybrid technique whereby points within a certain radius are grouped together and indicated with circles that contain a number that denotes the number of point features in that location. For now, these are called *marker clusters*, and implemented within Leaflet, a JavaScript library, though other technologies will probably implement their own methods for their creation in the future. This technique is particularly helpful for quickly visualizing data comparisons at small scales. Another advanced technique that can be useful as an alternative to choropleth mapping is hexagonal binning. In certain software, a regular grid of hexagons can be created from data that would otherwise be obscured by the units that organize the data. For example, the BBC News uses an equal-area hexagon grid, in a yellow-red-blue color scheme, to explain the history of elections in the United Kingdom. Since the map uses population-weighted statistics rather than area-weighted statistics, it makes more sense than displaying by counties.

Creating color gradients is thankfully fairly straightforward in most software products today. Built-in algorithms for stretching (interpolating) colors between a low and high value may be available. In CartoCSS you might simply provide percentages preceded by keywords such as "lighten" and "darken" next to the original color. Online tools can be useful in providing lightened and darkened colors in your color system of choice based on a single input color. Interactive color wheels can be useful in building multihue gradients.

What Color Are You Today? Color Connotations

Another color rule to become familiar with is that colors often carry with them certain connotations relating to feelings (sad, happy), conditions (good, poor), and things (trees, sky). Color connotations are not uniform across all cultures, but there still are some common map-related color connotations that do cross many cultural boundaries. These are listed in Figure 5.12.

Keep these in mind when designing maps so as to not send the wrong message. For example, trying to highlight wildlife corridors with a dark red hue, which in the United States is usually associated with danger or poor conditions, confuses the map reader if what you really want to convey is that they represent good conditions for wildlife. Another bad example would be the use of blue as a fill color for wildlife habitat polygons. The map reader can easily interpret those as bodies of water, even if you have a map key that states otherwise.

Please note that the purpose in discussing color connotations and in giving some examples of the common ones is to keep you from *blindly* disregarding

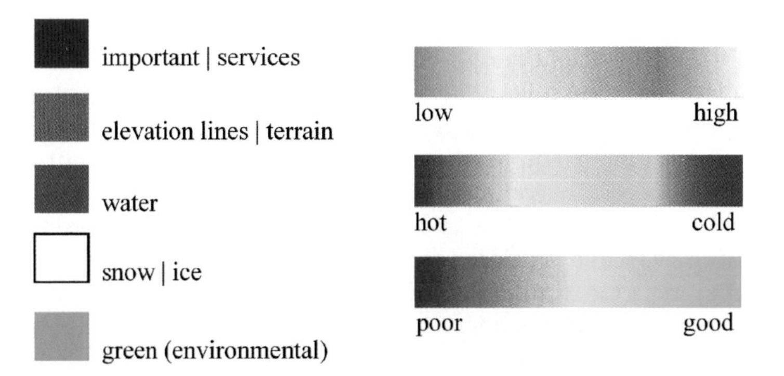

FIGURE 5.12
These are some color connotations that are applicable to mapped variables.

FIGURE 5.13
This elevation color scheme does not follow any of the usual low-to-high color ramping schemes. Because we are not used to seeing elevation depicted in this way, it has a surreal quality to it. To see these same data depicted with a typical low-to-high color ramp, see Figure 5.11.

conventions. However, these are subjective interpretations, not hard-and-fast rules, so if you can make a strong case for using them in an unconventional fashion, then by all means, give it a go. We ought not to feel constrained by convention, but rather guided by it.

Interestingly, color connotations are not very consistent across time periods, even relatively short time periods. For example, as recently as the early twentieth century, pink was a color associated with boys while blue was a color associated with girls. The previous edition of this book stated, "obviously the opposite is very much the mode now." Since that was written, it seems as though the color-fashion-gender connection may be waning again, as more males are now choosing to wear shades of pink, which just further proves the point that color fashion is in constant flux. When it comes to colors, a person's taste is both cultural and personal, and certainly not a unanimous doctrine.

One reason to rebel against the standard color conventions is to keep your audience on its toes and present them with something new, different, and exciting. For example, an interesting approach could be to change your colors from what might be a typical scheme to something a bit "off." An off color scheme can bring a science fiction feel to your work as in Figure 5.13, which is a rehash of Figure 5.11, but in more surprising colors.

As I hinted earlier, with the red wildlife corridor example, color connotations are not consistent across cultures. When you are designing for an international audience, if you aren't already familiar with the culture of the audience, you would do well to vet your color scheme with someone native to that culture. One example of a color faux pas that could occur involves the use of the relatively common green-to-red color scheme that is usually used on a map to depict the low-to-high values of a variable, where the "bad" amount of the variable is colored red and the "good" amount of the variable is colored green. One example of this might be a business map showing the good places to increase business operations as green and the poor places to increase business operations as red. In many European and American cultures, green evokes such things as likeability, spring, or environmental correctness, while red evokes such things as danger or poor environmental conditions. That seems to fit the data and the intended interpretations of the colors pretty well, but in China, red is the color of good luck and celebrations. The green-to-red color scheme may not, therefore, be a good fit for a business operation map that is intended for a Chinese audience. In this example and others, these differences in meaning could cause quite a communication mishap if your maps are color-cued to the wrong cultural audience.

Color Blending, or My Eyes Are Playing Tricks on Me!

As mentioned in the "Legend" section of Chapter 3, "Layouts," colors can be perceived differently depending on how they are combined with other colors. Therefore, I advise looking at many colors, in combination with those you have already chosen, before deciding on one for a particular part of the map or a particular feature. It can be very helpful to use a software tool to do this, or you can create a series of color combination boxes yourself on which to base your feature choices.[2] You could use something like the color boxes I put together in Figure 5.14, or you could simply change the colors within your GIS many times while looking at the end result. Reference boxes—these might be part of a style guide you put together—are often an easier way to go. The color boxes in Figure 5.14 show some color combinations that work well together. Perhaps you could create your own color combination boxes like these even when you don't have a particular project in the works, and keep them around for when you need a boost of color inspiration. The inspiration for this particular color set came from the spines of children's books that my kids had on their bookshelf.

Now, another reason to make color boxes like these is to get a feel for not just whether the colors clash, but also to figure out if they play any eye tricks on you when they are overlaid. Some colors, when they are superimposed, create certain visual effects that aren't desired on your map. When it comes to GIS maps, these effects are not as noticeable as they are to those in other graphic design fields, but it is still important to have a cursory understanding of what they are and what causes them. These are discussed below.

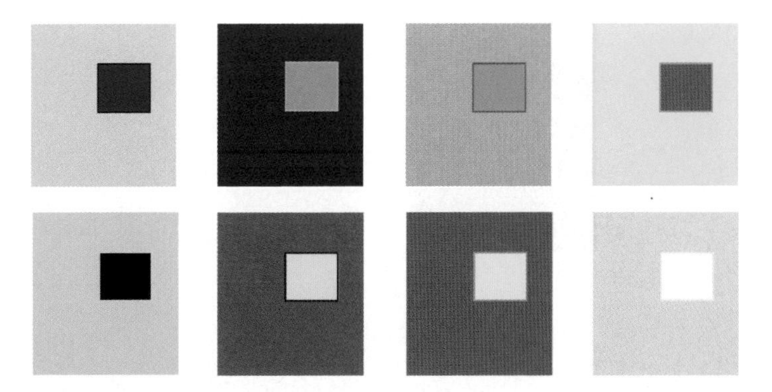

FIGURE 5.14
These boxes show a series of color combinations that I found by looking at various book covers and spines on children's books. These color blends were already vetted by the graphic artists at the publishing houses, giving the blends some credibility. Look around you and you will likely find many sources for color inspiration.

FIGURE 5.15
The color combination on the left can make you dizzy, as it appears that the red is jumping out at you. The color combination on the right makes the blue look as if it were receding away from you.

First, there are cool colors and warm colors. Go back to the color wheel in Figure 5.2 and look at all the colors on the left-hand side: blue, green, and purple. Those are cool colors. The colors on the right-hand side are the warm colors: red, yellow, and orange. When warm colors are placed on top of a cool background, they will jump out at you, creating a stereoscopic effect, and perhaps make you dizzy. If your layout has a blue background, for example, you won't want to print your title text in red. Conversely, when cool colors are placed on a warm color background, they can look as if they are receding away from you (see Figure 5.15).

Second, there is the situation where the same color can look different depending on the color saturation of the background color. If the background is dark, then the foreground color will look more vibrant; if the background is light, then the foreground color will look less vibrant (see Figure 5.16).

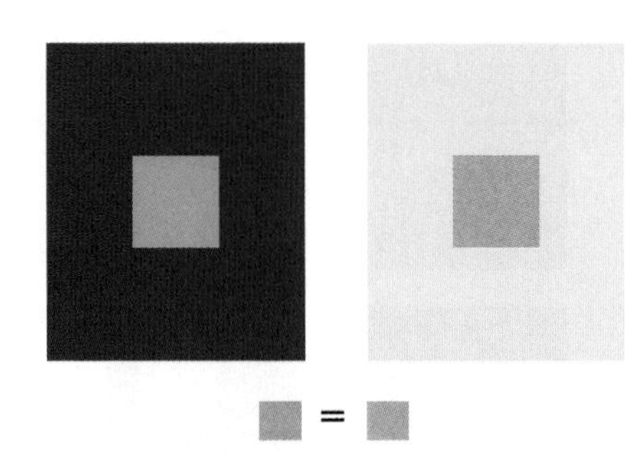

FIGURE 5.16
Both shades of orange are exactly the same in this example. The one on the left, however, looks more vibrant than the one on the right simply due to a change in the darkness of the background color.

Contrast

Another visual perception concept that is of significance to GIS is color contrast. High contrast between features on the map element, especially, is important because without it you risk obscuring the salient points of the map. The purpose of contrast is to present features so the map viewer can easily distinguish them from one another, and more importantly, from the background. For example, a typical paper subway map shows the main subway line features in highly saturated colors like blue and red, while the background is pure white. The viewer's attention is focused squarely on the subway lines, as it should be, because they are the main purpose of the map. The high contrast enables the small subway line features to be clearly seen. Navigational maps also exhibit high contrast between the background, basemap, features, and the features of interest. For example, a community group might give out maps of hospitals and emergency clinics that show the emergency facilities as vibrant red squares on a pale basemap so they are clearly visible to those who may very well be in a rush.

A word of caution: novice cartographers often make the mistake of creating maps with too many saturated colors or not enough saturated colors. Some people are afraid to put too much color on a map for fear that it will stand out (what a problem!), and conversely, some seem to try and mask a deficiency such as lack of design skill or lack of confidence in the map's purpose by blanketing the map and layout in bright colors. Whatever the reason, the novice generally requires much experimentation, mindfulness of color contrast, and critical feedback to improve this situation.

While the subway map example and emergency services maps illustrate good contrast by means of a few dark colors on an otherwise subdued background, it is also possible to create a nice map with the opposite type of contrast: a dark background with lighter focal features. For example, let's take a look at the Spain maps shown earlier in this section (Figure 5.10). My point with those was to point out the five-shades rule, so I didn't include any of

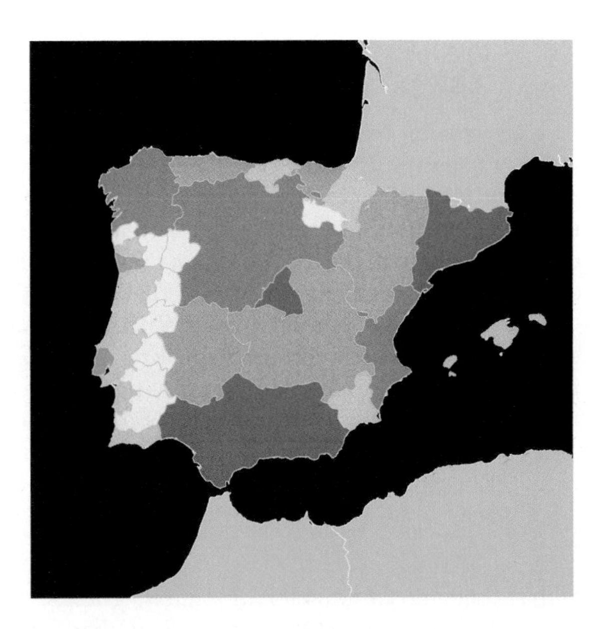

FIGURE 5.17
The map shown here exhibits a type of contrast (dark background, light foreground) that is the opposite of most people's idea of good map contrast (light background, dark foreground), but the bold colors make it a great graphic for a report cover or website graphic.

the surrounding geography as it was unnecessary for the discussion. But if we retooled the map for a different purpose, such as a website graphic or a report, we may want to add background information such as the surrounding country boundaries and water features to provide more geographic context. In doing so, we could make the added background features very dark, with colors such as olive green and navy blue. In doing so, we could make the added background features very dark, with colors such as olive green and navy blue, to match the original figure's light on dark color scheme. The resulting map (see Figure 5.17) illustrates the high level of contrast the dark background provides against the light peach color of the main features (Spain and Portugal) and additionally exemplifies a bold, eye-catching graphic style. This light-on-dark type of contrast has exploded in popularity since the first edition of this book was printed.

In summary, if you take care to provide adequate contrast in your color choices, whether it is a dark background with lighter main features or a light background with darker main features, you will be sure that the most important data are brought to the viewer's attention.

Color Everywhere Is Tiring

We've previously focused a lot on color rules for the map element of a layout. Now I want to turn your attention to some of the margin elements of a layout

EXPORT FILES AND COLOR

Exporting to compression-based file types like JPEGs and GIFs can cause your colors to change. In some cases the change is just to a similar color, but in other instances the color can change dramatically. The compression algorithms themselves are to blame for these color changes because in order to make the file smaller (i.e., compressed) they lose some of their color specificity. With GIF file types we generally have more control over these potential color shifts because, depending on the software from which you are exporting, some settings can be changed to deal with the issue. However, with JPEGs you cannot minimize the color-shifting yourself unless you change the compression (sometimes called *quality*) of the file. Sometimes this is an option and sometimes it is not, depending on your software.

that can themselves become quite colorful: charts, graphs, north arrows, menu bars, and icons of any sort. The thing to remember for these margin elements is that while it is tempting to "decorate" them with many colors, it is often unnecessary to do so and can instead grab attention away from the map element. The following excerpt from an article by Stephen Few illustrates this point: "conferences dedicated to color and its use for information displays, exhibit great restraint in their use of color during presentations. For example, their PowerPoint slides tend to include grays far more often than you'll typically see elsewhere. These experts realize that color should be used meaningfully, not arbitrarily or gratuitously."[3]

Now, if you are still keen to use color in your layout margins, it is imperative that you put as much thought into those colors as you did for the map colors. For example, I mentioned earlier that you can borrow a color used in the map by coordinating the title or layout background with the same color used to highlight features in the map. This is called *color echoing*. It is analogous to the poster finishing technique where the framer chooses frame colors by identifying important but sparse colors in the artwork. Similar to the framer's goal, we choose a color that will complement the map but not compete with it. In closing, if you are unsure whether the color is distracting and superfluous or complementary and meaningful, you would be wise to leave it out.

Ensuring Readability for Color Vision Deficiency

At the risk of sounding like a broken record: you must consider your audience when you design. One of the biggest audience-related design issues that mapmakers continue to neglect is designing for those who have a color vision deficiency (this is more colloquially called color-blindness). What a waste of time it is for everyone when you come up with a glorious color

> If you create a map for your workgroup, and your workgroup consists of three Caucasian men, the chance that one of them is red-green color deficient is about 22%.

scheme only to later find out that your boss, who asked for the map, can't tell the difference between the red counties and the green counties. Taking this a step further, let's say you designed a map of park trails for the public, but then you find out that 10% of the public can't tell the difference between the colors for the horse trails and the colors for the hiking trails. Not only have you wasted many people's time, but you have also created a communications disaster.

It is in every geoprofessional's best interest to have an understanding of what color deficiency is, how to design around it, and how to check a map to make sure it is readable. So first let's focus on what color deficiency is. The term describes a whole range of deficiencies that range from mild inability to distinguish a few colors to the very rare inability to see any colors except black and white. The most common deficiency is red-green, which causes both red and green colors to appear brown. Red-green deficiency is (usually) a genetic condition that affects about 8% of men and 0.5% of women of Northern European descent. The rates are much lower in other populations: about 5% of Asian males and about 4% of African males, for example. A rarer deficiency is the inability to distinguish between blue and yellow. The blue-yellow deficiency appears in the same proportion of men as women, but affects only about 0.01% of all people. The rarest deficiency is black-white, which affects only about 0.003% of all people.

How do we design our maps to accommodate color-deficient individuals? It is important to note that with red-green and blue-yellow deficiencies, colors can still be seen, but not as the colors we intend. One of the biggest problems with GIS design when it comes to this issue is when a viewer can't distinguish between the colors on a color gradient map (choropleth or heat map). Indeed, a very common color ramp is the green-to-red ramp, and this, in particular, can be useless for those with red-green deficiency.

This is especially true when the polygons, points, or lines are numerous and close together. In some cases, maps with only a few large red polygons and a few large green polygons, for example, are still distinguishable by red-green color-deficient individuals. Perhaps these test subjects only have a mild form of the deficiency, but there is also a size issue here. Ishihara hidden-digit plates, which are commonly used to test for color-blindness, contain many quite small, multicolored dots with a number hidden among them. Those who are color-deficient can't see the number and those who aren't color-deficient can see the number. The fact that the test requires the small dots of color points toward the fact that it is the number of and size of the colored dots that matters in addition to the colors themselves.

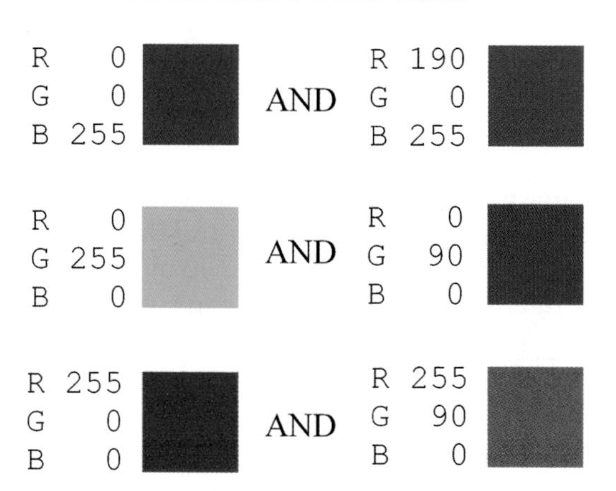

FIGURE 5.18
When choosing colors for map features that will be near one another and need to be distinguishable by red-green color-deficient individuals, avoid color pairs that are only different in their red and green values.

When designing, be wary of using colors that are low in saturation because they are hard to tell apart. A common map technique, for example, is to use a pale blue for water features and pale yellow for land features. These two colors are a lot harder for blue-yellow color-deficient individuals to tell apart than high-saturation colors. Another situation to be on the alert for when designing for red-green color-deficient individuals is using two colors that are the same for all components except green and red. A blue and a purple, for example, which differ only in their red component, may appear the same (see Figure 5.18).

Additionally, reds and greens that are pure red and pure green are the biggest problem makers. When a red is tinted somewhat with, say, some blue, and a green is tinted somewhat with some yellow, they become easier to distinguish from one another. Therefore, if you take care to provide sufficient contrast in your colors you might be able to circumvent the color-deficiency issue. This can also be accomplished by adding differences in brightness between the colors. Although the color-deficient individuals will not see the colors the same way other people do, they will at least be able to tell the difference between the map features, find them on the legend, and correctly interpret the map.

When do you need to pay attention to color-deficiency issues? If you ever design a map for anyone other than yourself (supposing that you are not color-deficient) or your immediate workgroup, it would be wise to design it as if someone with a color-deficiency were going to view your map. Certainly on maps that will be widely distributed or shown at a conference, you will want to ensure that the colors are visible to *all* audiences, including the colorblind, by running it through a color-deficiency simulation algorithm, such as Vischeck.[4] This is a type of website where you can upload a map and have it transform it into a graphic that shows you what your map would look like to a color-deficient person. Another option is to begin with a color palette that is specifically designated as visible to color-deficient individuals. The ColorBrewer website is a handy reference for such palettes.[5]

Inspiration

While the inspiration section is the last of the three Circle of Color Love concepts to be discussed, it is not the least noteworthy. In fact, it is here at the end of the other two sections as a means of giving those other concepts some attention since it could just happen that once you read this section you may not feel a need for the other two. I say that because the shortcut method to figuring out how to make a beautiful map, at least where color is concerned, is to borrow the color scheme from an inspiration piece. This is what I call the *interior design* method of color choice. Have you ever seen a TV design show where the designer reveals that the entire room's tone, style, and color were chosen based on one measly painting they found in the homeowner's shed? Well, the inspiration piece and its initial location may vary, but you see the point. Yes, you too can make derivations of other people's work, and the big secret is that all designers do this, even map designers.

In practical terms, the "inspiration" method of choosing colors for a map and layout is this: next time you are looking at a nice piece of art online, in a magazine, newspaper, or museum, take a moment to mentally (or physically) record the color scheme in as much detail as possible. What colors are used on the focal point? What colors are used in the background? How do they blend together? What kind of feeling do the colors relate to you? You can then adapt those colors to your map and layout. Yes, what I am saying is, let other people do the work of choosing colors that look good together for you!

If classic art isn't your thing, then look for inspiration in other places. How about fabric? Speaking of, mid- to high-end retail clothing companies often go to great lengths to produce catalogs and websites to showcase their wares, which can give you hints as to which colors are currently in fashion and how to put them together. As I stated earlier, I have used book covers and spines for color pair ideas. Yet another option is to get some painter's pamphlets (you'll find many free ones at paint shops) that show colors in groups of three or four coordinated colors. The possibilities are numerous, obviously, and it just takes a little creative thinking and a watchful eye to discover them. Some potential sources are enumerated here:

- Nature (photographs or real-life excursions)
- Fabric swatches (from your own furniture or clothing or a fabric store)
- Clothing catalogs (online or print)

- Book covers and spines
- Painter's leaflets (get online or at a paint store)
- House colors (cruise the neighborhood)
- Paintings (art at museums or online)
- Other people's maps
- Flower arrangements (look at florist's examples on the web)

Once a general color palette is chosen based on an inspiration piece, you will have to adapt it to your map's purpose. That's where your knowledge of color theory and rules comes in. In some cases you may wind up using the exact colors from your inspiration piece, but usually you will find that as you build the map and take its particular nuances into account, it will veer away from a literal translation of the inspiration piece and wind up standing out on its own. And then, for the ultimate compliment, perhaps your map will serve as an inspiration piece for other designers.

Grayscale Modern

Now that the color basics have been laid out, it's time to retreat a little and say something about black-and-white maps. When the field of GIS first started, black and white—grayscale—was the norm for most maps. It was simply too laborious and expensive to create color maps with user interfaces being either nonexistent or very simplistic and color printers hard to come by. Now, of course, color is ubiquitous because we have easy tools in our GIS to modify colors in myriad ways and because they are easy to display on our modern devices and cheap to print on our office's color printers and plotters. So because of this near-universal use of color, it is actually difficult to come by a map that only has a grayscale palette. The only exception to this might be a map published in the interior pages of a printed newspaper. So why might someone consider using grayscale in a modern map?

First, a resurgence of grayscale maps might be in the making as people start to backlash against the use of too many colors, particularly the use of colors that aren't even effectively communicating anything. Grayscale forces you to communicate elegantly and succinctly. Use color when it is necessary but always at least consider grayscale when you are designing for a map that does not contain a lot of variables to differentiate.

One example of this is the *New York Times* graphics department style. Maps that are created through their graphics department for print or for digital devices typically have a common color scheme that can best be described as subdued. The color palette has a lot of grays, with color being reserved for the most important data overlay. Picture a grayscale background with lightly saturated green and red dots to denote population density, for example. These maps are widely regarded to be some of the most elegant and informative in the profession. They represent a large cartographic style difference from the usual GIS map style of a decade ago.

Another reason a geoprofessional might turn to the neutral side has to do with confidence. As the geoprofessional becomes more confident in design, he realizes that color is not the absolute necessity that he once thought it was. His skills at placing objects on the page in appealing ways and coordinating line thicknesses and other noncolor aspects of the design might make him want to ease up a bit on color in order to allow those other design elements to work effectively. In this way, the color is lessened so as to be less intrusive on the design. And finally, yet another reason for using a neutral scheme is simply that we want to bring variety to our map products, and what better way than to surprise people with an elegant and tasteful grayscale map. In a time when people have access to color at every turn, these maps can be refreshing, surprising, and attention-grabbing.

Endnotes

1. H. Meihoefer, "The Utility of the Circle as an Effective Cartographic Symbol," *Cartographica: The International Journal for Geographic Information and Geovisualization* 6, no. 2 (1969): 105–117.
2. One such tool is ColorBrewer developed by Cynthia Brewer, professor and associate head of the Department of Geography, Pennsylvania State University. The tool can be found at http://www.colorbrewer.org.
3. S. Few, "Practical Rules for Using Color in Charts," *Perceptual Edge Visual Business Intelligence Newsletter* (2008), http://www.perceptualedge.com/library.php (accessed October 8, 2013).
4. B. Dougherty and A. Wade. Vischeck website, 2008, http://www.vischeck.com (accessed October 8, 2013).
5. ColorBrewer, http://www.colorbrewer.org.
6. D. P. Finlayson. "Combined Bathymetry and Topography of the Puget Lowland," 2005, University of Washington, http://www.ocean.washington.edu/data/pugetsound (accessed October 8, 2013).

Resources

4096 Color Wheel is an online tool for picking colors. It gives you a range of colors on a color wheel to choose from, and once one is chosen you can modify its properties and output a hex value for the color you create. http://www.ficml.org/jemimap/style/color/wheel.html.

Adobe's Illustrator/About color page presents information explaining the RGB, CMYK, HSB, and Lab color models. http://help.adobe.com/en_US/illustrator/cs/using/WS714a382cdf7d304e7e07d0100196cbc5f-6295a.html.

CIELAB Australia has a great visualization of the CIELAB color model on its site. http://cielab.com.au/?p = 66.

Color Scheme Generator 2 is an online tool that allows you to interactively choose a color palette using a clickable color wheel. Analogous, complementary, and polychrome palettes are created based on a single color choice by the user. http://colorschemedesigner.com/.

Color Wheel Pro, TiGERcolor, and **Genopal** are software programs that allow the creation of custom colors and palettes. http://www.color-wheel-pro.com; http://www.tigercolor.com; http://www.genopal.com.

COLOURlovers is an online community where members post colors (individual or in palettes) and patterns, paired with a ranking system and critical feedback from other members. Informative articles and trend-spotting features are also available. This is a great place to go during the stage when you are picking out colors and need specific RGB triplets to go with whatever you find. http://www.colourlovers.com. Also see their article titled "Common Color Names for Easy Reference," http://www.colourlovers.com/blog/2007/07/24/32-common-color-names-for-easy-reference/.

Kuler, by Adobe, is another community color site that allows you to create color palettes. http://kuler.adobe.com/#.

Perry-Castaneda Library Map Collection provides many maps that can aid in color picking and inspiration. http://www.lib.utexas.edu/maps.

Strange Maps is a widely read blog with a variety of maps and corresponding critiques. These maps are useful to glean ideas on novel approaches and color schemes. http://bigthink.com/blogs/strange-maps.

The Code Side of Color is an illustrated, well-written primer on the hex color system written by Ben Gremillion for *Smashing Magazine*, 2012. http://coding.smashingmagazine.com/2012/10/04/the-code-side-of-color/.

Transit Maps of the World is a comprehensive collection of transit maps that is a great resource for color, especially if your map is as complex as some of the ones shown in the book. The book also sports good design in its own right and can thus give you ideas for how to incorporate your maps into a layout, report, or book. Mark Ovenden and Mike Ashworth, *Transit Maps of the World* (New York: Penguin Books, 2007).

Study Questions

1. List and describe four types of color harmonies.
2. Describe the hex color system and its use.
3. Why does figure–ground matter? Find three maps with good figure–ground differentiation and describe them in terms of color contrast, which features are the figure and which are the ground, and what colors and techniques were used to achieve this.
4. Choropleth maps, heat maps, and elevation maps use color gradients as their primary means of communication to the map reader. What is a choropleth map and what usually needs to be done to the data before applying a color gradient to it?
5. Explain the difference between choropleth maps and heat maps.
6. What is red usually used to connote in Western countries? Blue? Green?
7. What happens when a small orange feature is placed on a purple background?
8. What is color echoing?
9. Why must a cartographer care about color vision deficiency? Cite three details pertaining to color vision deficiency as it relates specifically to cartography.
10. What is hypsometric tinting?

Exercises

1. Imagine that you own a cartography company. Create a book of shop colors with five color palettes comprised of four colors each. Showcase the color palette by including a color swatch for each color along with its corresponding code in hex and RGB, naming it, and including a simple map using all four colors. The map should use Natural Earth data or some other easy-to-access data source, and can be of any location the student wishes as long as all four colors are used in the map. Use exactly the same map for each color palette, only swapping out the colors. Hint: get color palettes from inspiration pieces.
2. Using the same map that you created for the book of shop colors in the previous exercise, create a clear figure–ground differentiation using either a halo or a vignette along with a subdued color coupled with a strong color. Provide a short description of how you created it.

6

Features

This chapter presents mapping standards and techniques for many common geographic information system (GIS) features. These feature discussions are also a springboard for mapping techniques that can be used for GIS features that aren't listed here, since many of the techniques are widely applicable. Please note that these are all just suggestions, and you must take the responsibility for applying them appropriately to your map, given that it has its own unique purpose, data, and circumstances. For example, in the "Bodies of Water" section, the color palette shows eight shades of blue as suggestions for water feature colors. While blue is a standard color choice for these features, you might want to make an exception for a particular map. Perhaps you feel that light gray would be good for the lakes on your map because you are required to use shades of blue for a different feature. In this case, bucking convention allows the lakes to be differentiated from the other feature type, a good reason for changing things up. Some of the ideas in this chapter encourage creative solutions by giving examples of some innovative ideas. But do not let those limit your creativity. Come up with your own original solutions too. By learning the conventions, you should become confident in determining where there is some room for creativity and where you want to stick with the traditional methods.

Roads

Colors*

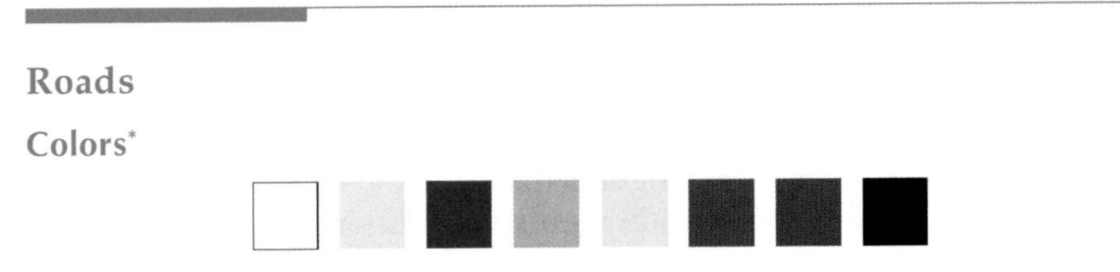

Roads, also called arterials or motorways, are fussy features. They can include everything from the major interstate highway to the dirt road behind your house. A map showing all the road types from dirt to autobahn with the same color and line width, at any scale, would simply look ridiculous unless your goal was to produce a jumble of lines (see shaded box, this section). There are exceptions to every rule, but the for the most part our roads must be separated into size classes and symbolized accordingly. The color bar above represents suggested map colors.

* The color bar contains suggested colors for roads.

Once a symbol is chosen for each road class, with a thinner line for minor roads and a thicker line for major roads, the drawing order must be addressed. A map is more comprehensible with the smaller roads drawn underneath the larger. Furthermore, when it comes to major intersections with all kinds of on and off ramps cluttering up the picture, drawing order should reflect reality. That is, if one ramp goes *under* another in real life, it should show up as being under the other on the map.

Once thicknesses and drawing order are sorted out, you can focus on style. If the road is symbolized with a line edge, otherwise known as a *cased line*, it is broken at intersections to provide visual continuity. Cased symbols are usually drawn with a light color in the interior of the road line feature and a darker color for the outline of the road line feature. If this rule is reversed, the road will not look as if it has width or dimension, and instead will appear as if it were highlighted. Road labels are placed inside a cased symbol. A darker line within the casing can provide center-line realism. Casing is not recommended for scales above 1:100,000 (for these, use a regular thin line for major arterials). Directionals (arrows) can also appear within the line and are usually used for major arterials or one-way city streets. Special road labels for major motorways are called *road shields*. These typically look the same as the road shields you would see on highway signs, but they can be more generic. When symbolizing roads with road shields, you must find the appropriate shield file (many are found on Wikimedia) and determine how to label it. Road data for this purpose should have a name length field that you can use to select the appropriate shield size depending on the length of the name. For example, I-25 will need a larger road shield than I-5. If styling your data via code, it is best to find a code example for this and modify it as needed.

You will also have to be mindful of generalizing roads depending on the scale of your map. For medium- and small-scale maps, you will find yourself getting rid of the smaller roads in order to neaten up the finished map. This is easy to do if your data is already separated into road classes either via separate files for each class type or via an attribute that holds the class information. If your road data does not contain this information, a work-around that often does the trick is to calculate the length of the roads and then separate out only the longest roads. The longest roads are often the major roads. You will have

OUT OF THE BOX

If roads are the focus of your map, you could bend the rules. For instance, a small-scale picture illustrating all roads in one symbol type could make a statement about population, interconnectedness, or landscape modification. Another possibility is to create an oblique map of roads overlaid onto topography, but toggle the topography layer off to visually compare relative slope.

MapQuest.com used to be one of the most used online road mapping websites. Prior to mid-2006, MapQuest's online road map showed nonmajor roads as solid lines of a single color and narrow width. Google Maps came along at about this time with an online road map that used casing as the primary symbology for all of its roads. These maps looked much better and were easier to understand than MapQuest's, and it wasn't long until MapQuest followed suit by creating cased lines on their maps as well.

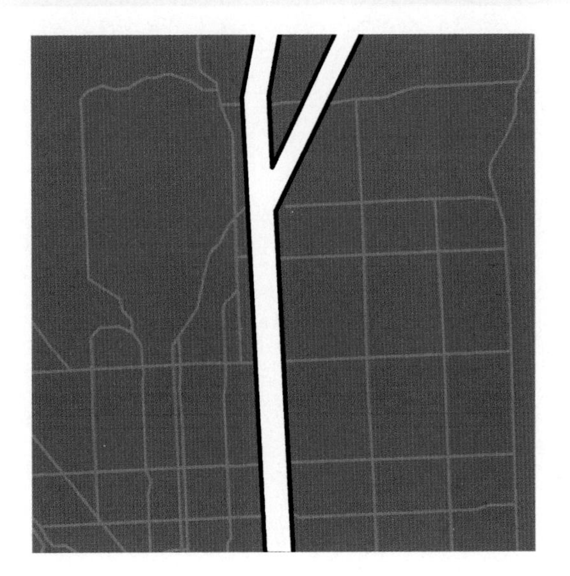

FIGURE 6.1
The highway feature is shown with casing, while the minor roads are shown as simple lines. Note how the casing doesn't interfere with the highway split.

to use trial and error to find out if this will work for you. Figures 6.1 through 6.4 are examples of how to depict this feature in certain situations.

Rivers and Streams

Colors*

Rivers and streams, also called *hydrography,* can be simple features to display if you adhere to some common conventions such as the almost exclusive

* The color bar contains suggested colors for rivers and streams.

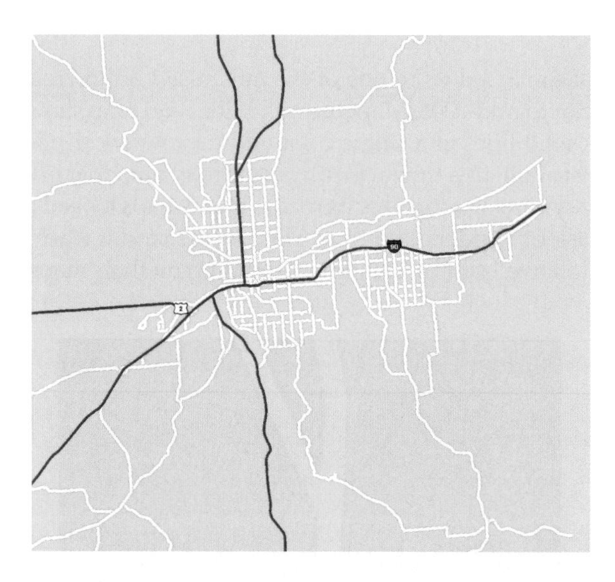

FIGURE 6.2
A classic land background is shown: green with a hint of brown (RGB: 209 224 115). The minor roads are white with gray casing. Solid red lines form the main arterials.

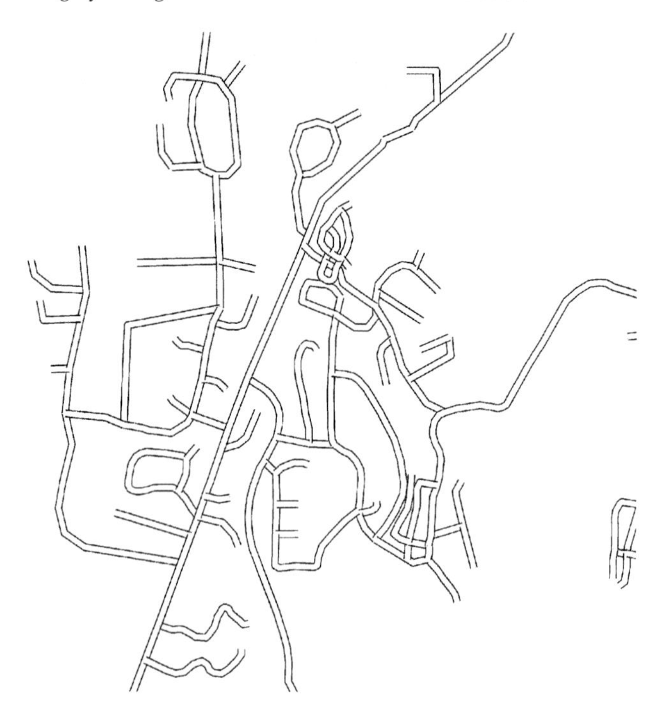

FIGURE 6.3
Evoking an architect's plan, this white background with white roads encased in black is suitable for professional black-and-white maps at a neighborhood scale.

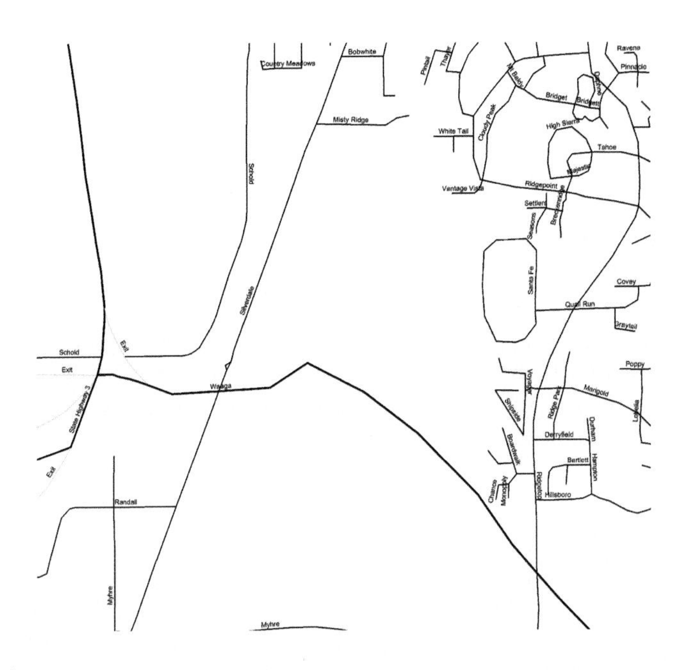

FIGURE 6.4
Shown here is a no-nonsense conventional road map with black lines in varying thicknesses according to road type. The white background and black road labels make it an interesting read at both neighborhood and regional scales.

use of blue to depict them and the use of lines instead of polygons for most maps. Therefore, you can simply slap on a line-based river dataset, color it blue, and go! No, it can't be that easy! Here are some additional considerations. To start, in many cases it is wise to match the stream line colors with the same color you are using for other water bodies on your map, including oceans and lakes. Other things to think about are how to show hierarchy, attributes, and labels. These are discussed next.

Showing Hierarchy

Rivers and streams in the natural world participate in a hierarchy determined by the volume of water they contain. The smallest streams in terms of volume, width, and depth, are the headwater streams that do not have any tributaries flowing into them. The second smallest is a stream that has only headwater streams flowing into it. At the opposite end of the spectrum are the largest rivers that flow only into large bodies of saltwater. These varying degrees of stream size can either be ignored on the map, by simply symbolizing them all with the same line width, color, and style, or they can be emphasized via different line widths, colors, and styles.

The display of hierarchy can become as complicated as the number of orders your streams represent. If you only have three stream orders, then a simple gradient from dark blue for the mainstems to light blue for the headwater tributaries may suffice. Furthermore, line widths can be tied to the stream order by using a thicker line for the mainstems and thinner line for the headwater tributaries. The hue gradient and line thickness techniques could be used separately or in tandem. To add yet another level of complexity to the stream symbology, the ephemeral streams can be depicted via dashed or dotted lines to emphasize their transient nature. All of these special cartographic effects are only possible, of course, if your stream dataset contains attributes pertaining to reach levels, hierarchies, or stream orders. If you don't have such a dataset, the next best thing is to assign the stream orders yourself by hand if there aren't very many, or find and run a stream ordering script (perhaps requiring a flow accumulation raster created from a digital elevation model) for your particular software.

Showing Attributes

Displaying stream attribute data, as opposed to simply the location of the stream (as above), can be extremely difficult depending on the scale of the map and the nature of the data. For example, a dataset where the stream has been segmented into 100-foot segments might contain an ordinal attribute, such as degree of channelization, which you want to display. A color gradient can be applied, but if the map is at too small of a scale, then you won't be able to see the color differences. To fix that, you might be tempted to increase the line width of a linear stream feature, but if the segments are short enough and numerous enough, you wind up with an incomprehensible mess (see Figure 6.5).

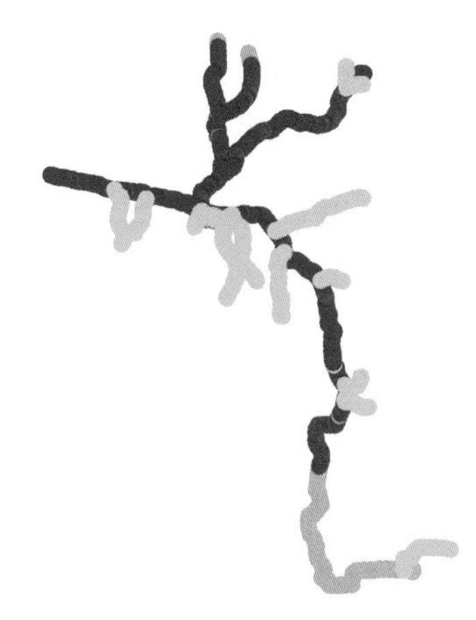

FIGURE 6.5
Stream segments in this example are color coded to represent an attribute value at each segment. Because the stream segments are so small compared to the amount of stream being shown on the map, the colors are squished and mostly incomprehensible.

Potential solutions for this predicament include generalizing the data so there are fewer segments, creating multiple maps at larger scales, and highlighting only segments that are above or below a certain threshold while keeping the remaining segments a neutral color. With some added effort, a hotspot map could be created by analyzing the data for clusters of high or low attribute values, delineating a circle around those values using the statistical results, then displaying the hotspots around the otherwise neutrally colored stream data. In this way you are essentially doing the viewer's work for them. Of course, if your data does not show true cluster trends, then this will not be the approach for you.

Labels

Stream labels are usually formatted in spline text, meaning the label follows the curve of the stream (see Figure 6.6).

Spline texts present all kinds of problems for the computer and therefore we often run into issues with fine-tuning the stream labels at map output time. Because spline texts are based on the curve of the stream, you won't want to create them until you know for sure at what scale the stream will be displayed, otherwise you run the risk of having super-curvy text paired with straighter line segments or super-straight text paired with very curvy line segments. If the spline text is dynamically linked to the feature scale, then your problems may be solved. Or not. In some cases, the spline text automatically generated is way too curvy due to the high number of vertices inherent in stream features. The resulting text can be difficult to read when every other letter is at a different angle. To ameliorate this, try simplifying the stream lines using your GIS and then basing the labels on the simplified stream lines. The labels can then be displayed with the nongeneralized data but will look somewhat better.

In addition to the splined text, another convention for stream labels is to use a script or italicized font in a blue hue (usually cyan, light blue) either slightly darker than or the same color as the stream feature itself. Depending on the scale of your map, you may need to increase the character spacing so that the label does not look as if it is squished in relation

FIGURE 6.6
This is an example of a stream label displayed with a spline text algorithm. Stream names should follow the curve of the stream they are labeling.

to the length of its associated stream. On the flip side, do not increase the character spacing to such an extent that you cannot easily see that all the letters belong to the same word! Also, not all of the streams and tributaries on your map need to have labels, even if your database contains all of the names. A type of map generalization is the paring down of labels to the number that are really necessary for the map viewer. Perhaps only the mainstems need to be labeled, or only the streams that are salmon-bearing, or only the ones in the central study area, for example. One additional note is to always include the stream name suffix (e.g., River or Creek) with the label. Also be sure to use initial capitals, also called *mixed case*.

Stream labels should always be placed above the stream lines, unless it is absolutely impossible to do so. Because English words have more ascenders than descenders, the label follows the line more closely and with more continuity if it is placed above the line. The favored direction for stream labels is toward the center of the map if possible. In some cases, for long streams and large map sizes, you may need to label streams more than once along their length. See Figures 6.7–6.10 for examples of how to depict this feature in certain situations.

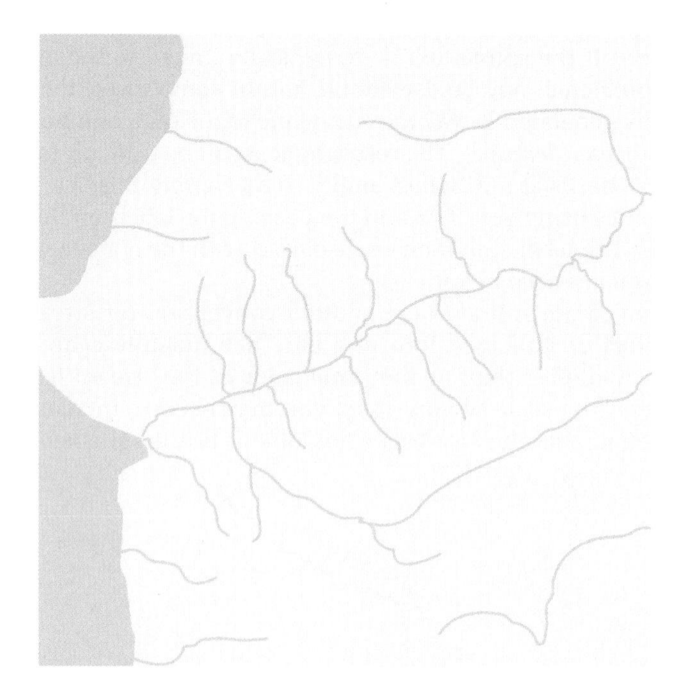

FIGURE 6.7
This is a basic stream map. All streams, regardless of stream order (hierarchy), are the same line thickness and color. The color of the stream lines matches the color of the outlet water.

FIGURE 6.8
These stream lines have two different colors and two different styles according to their stream order. The mainstem is a darker blue, the major tributaries are a lighter blue, and the ephemeral streams are thinner and dashed. The label text is splined and italicized, character spacing is wider than the default, space between "Rendsland" and "Creek" is also wider than normal. The label text color is the same as the mainstem color.

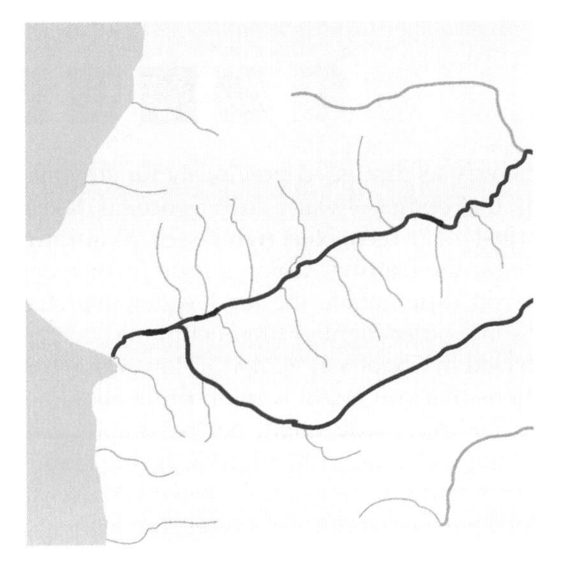

FIGURE 6.9
This stream map shows a few hot spots for a particular attribute (in this case it is large woody debris). The attribute is not depicted on all of the stream segments as that would have resulted in a cluttered and perhaps unreadable map.

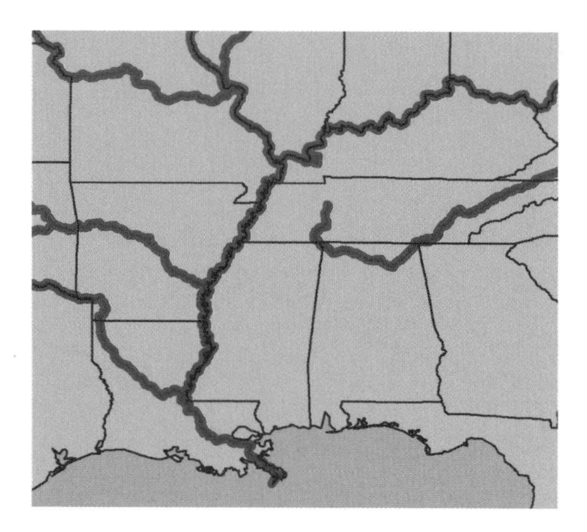

FIGURE 6.10
The rivers shown here are buffered to add emphasis and width to the stream lines. There is also a two-tone effect achieved by means of a slightly thicker river buffer (a light blue matching the Gulf of Mexico color) layered underneath a thinner, darker river buffer. Political boundaries that coincide with the rivers are layered on top of the rivers.

Bodies of Water

Colors*

In much the same way as discussed previously for streams and rivers, it is tempting to think that bodies of water such as oceans, lakes, and ponds are easy to style (you just make them blue, right?). Yes, conforming to convention means using blue for the features. Going a little further, convention would also dictate that you differentiate the land–water interface with a darker shade of blue or some other highlighting technique such as a halo, stipple, or vignette (described in Chapter 5, "Color"). However, what are some additional considerations that you might want to think about before going with the obvious? For one thing, you might not be simply trying to show the locations of your bodies of water; you might be trying to illustrate a variable associated with them instead. For example, you could use varying shades of blue and red to indicate temperature in a lakes dataset.

If you do go with blue for bodies of water data, let's discuss the shade of blue that you use. Is water an important part of the map's purpose? If so,

* The color bar contains suggested colors for bodies of water.

perhaps a dark shade of blue would look nice. What shade of blue goes well with your other features and background? For example, some maps show lakes and oceans as being so blue as to almost be black, and some do go so far as to display them as black features, in mimicry of aerial photography hues. Conversely, when water is an ancillary feature that will easily get in the way of the larger purpose of the map, the lightest color of blue is often the way to go.

You might also consider the texture of the water body. In many landscape architecture drawings, you will see changes in features highlighted with texture along the shared boundaries to provide figure–ground differentiation. In many cases the differentiation is created by simply displaying random dots along the boundary, in a stipple effect. In some historic maps, you will see the texture repeated about the entire water body as opposed to just the boundary. In this case, the dots or splotches evoke breaking waves at the surface. This is akin to displaying mountains with little upside down Vs or triangles. While these are unconventional approaches in the traditional GIS world, digital mapping products are now producing more maps with these kinds of effects as they incorporate more advanced image compositing capabilities.

Do you want to show the bathymetry (depth) variations? These can add a lot of interest to a map even if they are not central to the main point (see Figure 6.11). Why would this be? For one thing, the presentation of unessential information—if visually appealing—gives the viewer a sense that this map is something important, something to be reckoned with! For another, the realism gives added contextual information that can actually make the map easier to understand and more quickly decipherable (see the section "Simplicity versus Complexity" in Chapter 3).

Labels

Water body labels, like stream labels, are often (but not always) written in a script font or italicized. Also like stream labels, they are often (but not always) written in a darker blue color than the underlying water body feature. If your water body is in a particularly dark shade of blue, then you may want to use

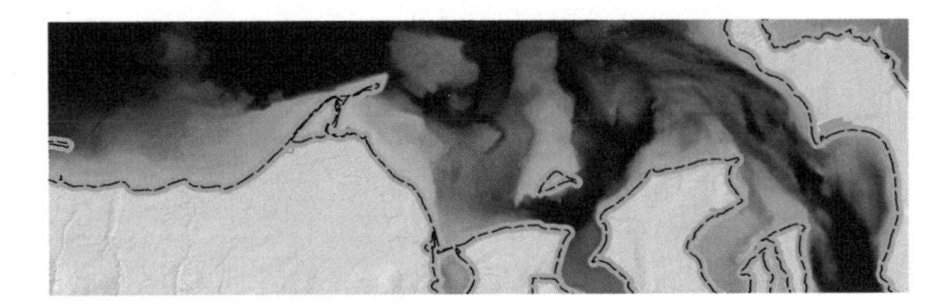

FIGURE 6.11
This map illustrates the visual appeal of bathymetric maps. The depth, displayed in varying shades of blue, adds a fluid richness to large water features.

a bold white text for the label. Now that we have the generalities out of the way, let's move on to the more complicated aspects of labeling. Please see Figures 6.12–6.16 for examples of how to depict this feature in certain situations.

You may find yourself, as I do quite frequently since much of my work involves Pacific Northwest data, mapping a region that has a particularly large amount of water to label. You don't just have an ocean and a bay, for example, you have an ocean, a strait, 10 bays, 10 inlets, a gulf, and an iceberg (well, maybe that's a stretch in the Pacific Northwest). You can either label just the major water bodies or label them all, but with different styles for each depending on the type. To get a grasp of how many different styles you

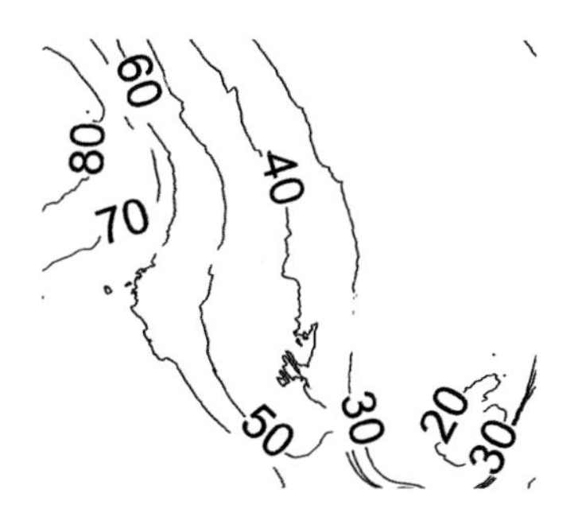

FIGURE 6.12
This is a straightforward bathymetric map primarily for use as a recreational boating and fishing aid. There are no extraneous details here and it is a very utilitarian map.

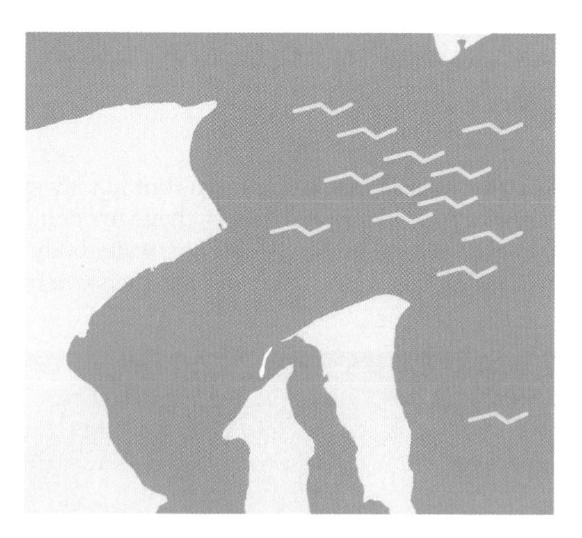

FIGURE 6.13
Here we have a modern map with an old-fashioned twist: wave crest graphics in the deeper parts of the water. They have no other meaning other than to further differentiate the water from the land (old-time cartography is notorious for superfluous graphics), but they are a nice cartographic effect when used in the right context.

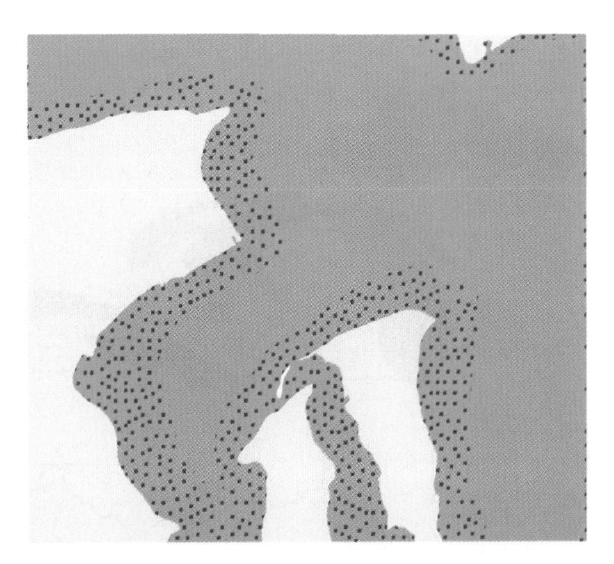

FIGURE 6.14
This is a map inspired by a landscape architecture-type drawing where stippling at the boundaries of two distinct features serves to accentuate the boundary and give some faux depth to the overall picture. The stippling could be a lighter or darker color than the background, whichever you prefer. Texture can be hard to come by on a typical GIS map, but this remedies that nicely.

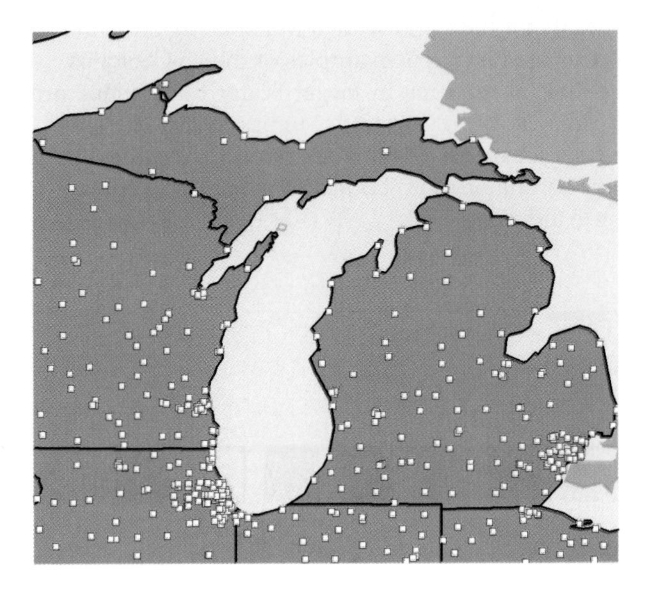

FIGURE 6.15
In this example, the lakes are kept a very light shade in order to keep them from overwhelming the map's larger purpose, which is to show the distribution of hospitals across the state of Michigan.

FIGURE 6.16
This dark blue results in real eye-popping emphasis and is much like what you might see in a satellite photo.

could use, look at a US Geological Survey (USGS) topographic map. A light blue (cyan) hue is generally preferred, and initial capitals are almost always used except in the case of oceans, which can be completely upper case. Swirly looking script fonts for bays and wide letter spacing on straits and other linear water features are just a few examples of the USGS styles.

Some of the special features in larger water body maps are everything from current arrows, buoy locations, temperature extremes, bathymetry (depth), and shoal locations, to name a few. Strict nautical maps, of course, focus primarily on one or more bodies of water and conform to strict standards in order to aid sailors.[1]

Cities and Towns

Colors*

Cities and towns, otherwise known as population centers, are stored as GIS data in either point or polygon form. The discussion begins with an examination of cities and towns in a point-data format.

* The color bar contains suggested colors for cities and towns.

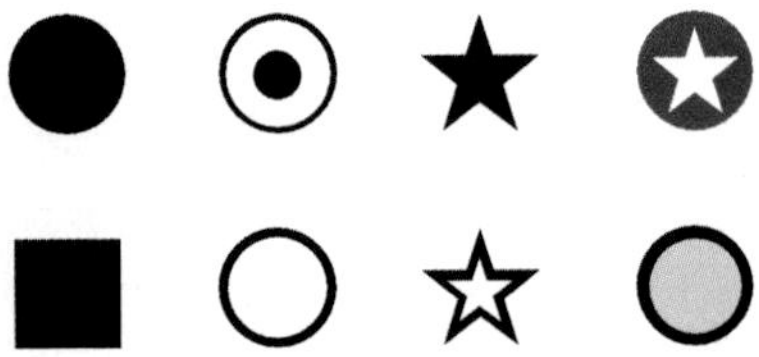

FIGURE 6.17
City and town points are depicted like the points shown here. Standards dictate that state and national capitals are symbolized with stars and circled stars.

Points

Our standard image of a city layer is of points corresponding to the location of the center of the cities, displayed as dots with city name labels near their associated dots. The points themselves, of course, can be graphically communicated in one of many different formats. Some of the most common point graphics for cities and towns include stars, dots, circles, squares, or some combination thereof. If the map shows a great many cities and towns, you will want to distinguish between the smaller and larger population centers by using multiple point sizes, multiple symbols, or both. This concept is referred to as *symbol levels*. An example of an effective combination would be a map that shows the major state capitals as stars, the smaller cities as medium-sized dots, and the large towns as small dots. This symbol level technique would be achieved in the GIS by means of a selection for population size and ensuing assignment of symbology for that particular population size (e.g., greater than 500,000) or by simply splitting the layer into multiple layers so that each may be rendered separately. The cartographic sophistication of your software will determine your methods in this case.

Colors are usually confined to primary or secondary colors (but not blue), with black and gray being the most popular, followed closely by white with a black border. A few examples of dot types and colors are shown in Figure 6.17.

Special Effects

When you have attributes associated with city and town point data that you want to display in some meaningful manner, a great trick to do this is via extruded graphics, especially when your display attribute is in ordinal format. For example, let's say you have some data on housing starts within a state. Each city and town within the state can be shown as a bar that comes out at the viewer via a three-dimensional (3D) effect at different heights depending on the number of housing starts recorded for that place in the past year. With enough data, the map can almost start to take on a continuous surface look and feel, like an elevation model of housing starts! A more typical example would be something like using the population of major US cities as the field for extrusion (see Figure 6.18).

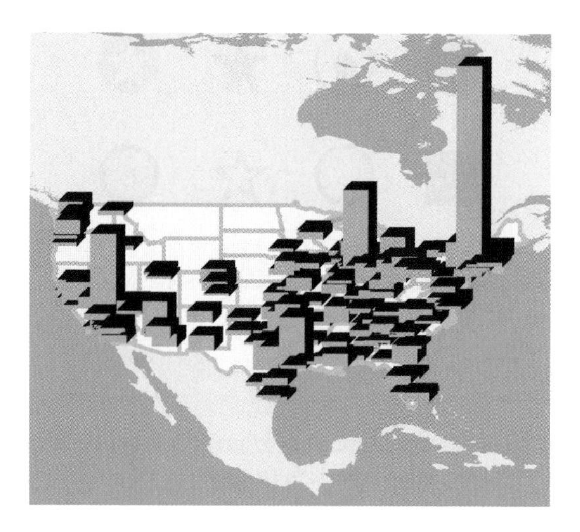

FIGURE 6.18
A variable is extruded so the viewer can make immediate inferences out of a large dataset. Shown here are the populations of major US cities.

Labels

Just as the points of a cities layer can be shown via different sizes and shapes of symbols depending on the size of the city, the labels can be varied for the same reason. State and country capitals, for example, are often labeled in uppercase lettering while the minor cities are left as mixed-case lettering. The use of bold fonts for the larger cities and regular fonts for the smaller cities can be used in addition to the case changes or independent of them. Yet another technique to further the distinction between city sizes is increased character spacing for selected cities. Labels are conventionally written in a sans-serif font, such as Arial, on modern-style maps. Label colors for cities and towns are almost always black or dark gray. The color of the labels might also be associated with some sort of attribute. For example, a map of drinking water contamination might use the color red for the five cities with the highest contamination levels. An additional item of note when it comes to city labels is that occasionally you might find yourself wondering why you need the point at all. If your audience only needs a general indication of where the city is or if the point would obscure some underlying data, you might choose to not display a point and instead, simply use the label as the geographic locator by placing the label directly over where the point would have been. This is acceptable as well.

Polygons

Polygon city and town boundaries (also called community boundaries) are common on medium- and large-scale maps, and are perhaps harder to show in a meaningful manner on a map than are points. With city polygons, you can easily obscure the other data you are trying to display by actually covering it up or by creating too much visual clutter. To fix this issue, you'll need to use a light translucent fill or only show the boundaries. The caveat for maps showing just the city and town boundaries without a fill is that the particular feature becomes less important to the overall map and, indeed, could simply be too hard to distinguish from roads, county boundaries, and other linear features. So your presentation of this feature will depend on how important the boundary is to the map's purpose, while being mindful to avoid too much clutter or the opposite, too little emphasis on the feature. Some options for these cases are as follows:

- Symbolize the polygons with a fill that is slightly darker than the surrounding land color while leaving the outline either minimal or absent (for maps without adjacent towns).
- For maps with adjacent towns, the fill color can be varied so that adjacent towns are distinct or a polygon outline with a slightly darker color than the fill can be used.
- Symbolize the polygons with a stippled, striped, or gradient fill to emphasize the areas while not completely obscuring the underlying details.
- Simply outline the boundary, and do not use a fill color.
- Create a transparent fill color.

Political Boundaries

Colors*

Many of the same design considerations explained in the polygon portion of the "Cities and Towns" section just prior to this are also applicable to the political boundary section. Political boundaries such as county, township, or country boundaries all delineate containment of land for political or administrative purposes and are stored as polygons or lines. Polygons are generally used for labeling, fill color, and boundary delineations, while political boundary lines are used exclusively for line symbology when a smaller dataset is desired and fill and labels are not needed. The political boundary datasets of the largest areas, representing accepted country boundaries as well as disputed boundaries, are often referred to as *admin 0* data. The smaller divisions within those countries, representing states, are often referred to as *admin 1* data. Seek out the most up-to-date political boundaries as they do change. Also be aware that there is more than one view of world

* The color bar contains suggested colors for political boundaries.

DASHED LINE PATTERNS

If you decide to put your political boundaries in a dashed-line format, be sure to convert them to line format prior to symbolizing. If you try to use a dashed-line format on a polygon outline, the dashes will be shown at varying intervals along the overlapping polygon outlines. So if you were using a dot-dash-dot-dash pattern on a polygon, you could wind up with a dash-dash-dot-dash pattern or something else entirely due to the overlapping effect.

administrative boundaries. In fact, some cartographers producing global commercial maps create three worldviews: World, China, and India.

Conventions

Always look up conventional styles for political boundaries since they are a good starting point. For example, county boundaries are often depicted as dashed or dot-dashed black lines over a gray bar. They are also often depicted with a simple black line, especially if the map is in a medium to small scale, showing many counties, for example. A map with just one or maybe up to three counties would be ideal for the dashed line technique.

Double-line boundaries are sometimes used to provide extra emphasis or differentiation from other boundaries. For example, if you have a nested zip code map, you can use single lines for the inside polygons that show the zip+5 boundaries, and use double lines for the larger zip code polygons. Thinner lines denote a subset of a larger feature when the larger feature is depicted with thick lines. Political boundaries, when they coincide with natural features, should be layered on top of the natural feature when possible. For example, country and state borders often follow major river courses, so the optimal overlap solution would be a thicker river course line underneath the thinner border line (see the "Rivers and Streams" section for an example). The borderlines running through the Great Lakes should be layered on top of the lake symbology.

Another hierarchical feature of political boundaries is the border between countries, between continents, and between countries and surrounding oceans. Which should come first? The ocean–land boundary ought to take precedence over the country boundary where they overlap. Indeed, this is another reason why a cartographer may wish to use admin 0 lines as opposed to polygons. With the line datasets, you typically only have the inland lines, not the ocean–land lines. Using these datasets enables you to symbolize the admin 0 boundaries on top of everything else *except* the ocean–land boundary.

A common map-making mistake is making a jumbled mess out of political boundary polygon fills and edge effects. For instance, let's say you need to create a thematic map depicting dominant crop types in a group of counties. Let's also say that you don't want to completely fill the county polygons because there is

The Four Color Theorem, which was proven by Wolfgang Haken and Kenneth Appel in 1976, states that a planar map of polygons needs only four colors such that no two adjacent polygons share the same color. This applies to only contiguous polygons, however, whereas political areas are often noncontiguous (e.g., Alaska and Hawaii vs. the rest of the United States).

elevation data you want to show under them. Trying to depict the counties with multiple-thickness outlines that are colored corresponding to their crop type is going to make for a disastrous outcome where shared boundaries need to be multiple colors and it quickly becomes impossible to read. Don't use multiple colors for the polygon outlines. Stick to a single, neutral color for those. Create interior buffers of the polygons such that each interior buffer can be assigned a different color based on the crop type. Ensure that the interior buffers are all the same width—don't use different widths or you'll end up with another mess. The polygon outlines, which are all the same color, go on top of the interior buffers. A very lightly saturated elevation map can be placed underneath everything.

Fuzzy Features

Colors*

A fuzzy feature is any feature that does not have an exact boundary or location in reality, while in your GIS database, it does. Sometimes a feature does not have an exact boundary for political reasons, and sometimes the exact boundary is simply unknown. Our problem with fuzzy features, as mapmakers, is that we have to store the locations in exact terms, but we ought to display them so that their lack of exactitude is readily apparent. This, for example, will assuage the fears of potentially affected people when overlaying certain data onto tax parcel data.

Let's take the example of a flood-prone zone analysis. You've spent weeks developing and computing flood-prone zones in a particular area to aid decisions concerning where potential environmental impacts might occur. Perhaps this analysis is all about pinpointing homes that should receive government assistance to update their septic tank systems for the good of the overall watershed. In this fictitious example, the people living in a flood-prone zone need to update their septic systems. You show your map, which is based on 30-meter resolution raster data overlaid on surveyor-accuracy

* The color bar contains suggested colors for fuzzy features.

parcel data, to a group of citizens, and everyone is trying to locate their house on the map to see if they are affected, when in reality, your map's purpose is just to show the general location of these flood-prone zones, not their absolute locations with respect to individual homes. When it comes down to deciding which homes will be affected, the flood-prone zone overlay will be used as a preliminary selection device to decide which homes will require on-site evaluations. One obvious way to avoid a rush of potentially needlessly concerned landowners is to avoid putting the parcel data on the map in the first place. Another is to point out the "fuzziness" of the data in the title, and yet another is to use some of the cartographic techniques mentioned here.

Technique

Fade-outs, squiggly lines, and dashed lines all convey a sense of transience that best suits the purpose rather than solid lines. Halos of decreasing prominence along the outer edges also fit the bill nicely. If a fill is in order, hash marks or stipples, paired with invisible outer edges, are good options since the outer edge of the feature will therefore contain a lot of white space and also convey some feeling of an unknown edge.

Of course, it should go without saying that whenever features with unknown boundaries like these are displayed on a map, especially if they are combined with high-resolution personal data such as parcel boundaries, a caveat needs to be made to this effect in the title, subtitle, or some other prominent location. If the graphics cannot convey the message, then the words can (and vice versa), but try for both. Please review Figures 6.19–6.30 for examples of how to depict this feature in certain situations.

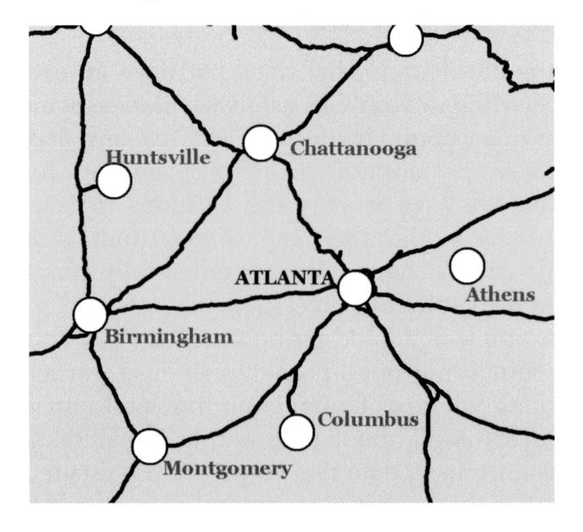

FIGURE 6.19
A very common and effective way to show city points is with white circles with black borders. The roads connecting the cities are also shown here, creating a central "hub" effect, with the roads looking like spokes coming off the hubs.

FIGURE 6.20
Town boundaries are shown here with a gradient fill, whereby the gradient gets darker as it approaches the edges of each town.

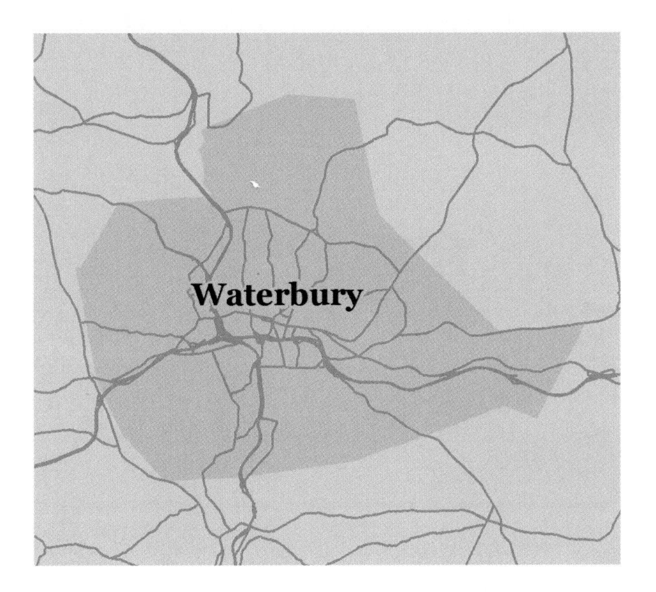

FIGURE 6.21
To emphasize just one of the towns without actually drawing a line around the boundary, this fill effect works well. To do this I simply reduced the values for each color component on the RGB scale by 15 (if you recall from Chapter 5, "Color," lowering an RGB number brings that color component closer to black). The background color is 215 215 158 and the Waterbury Township color is 200 200 143.

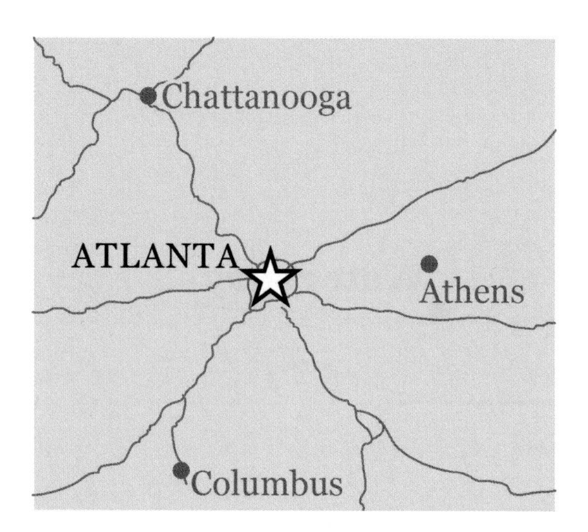

FIGURE 6.22
The city points in this example are differentiated by their symbol type and size, which correspond to their population. Stars are usually reserved for state or country capitals, and this holds true in this example for Atlanta, the capital of Georgia. The other cities are smaller circles.

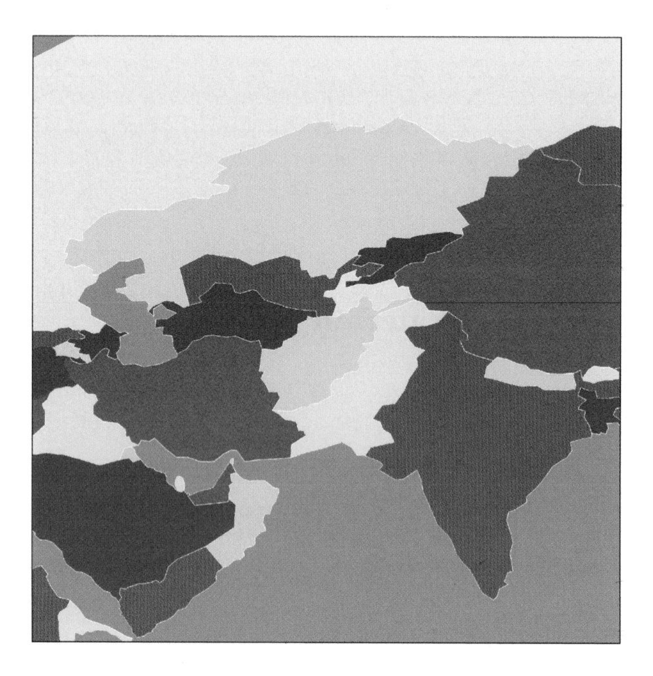

FIGURE 6.23
A very common way of separating political boundaries is to use different fill colors for each adjacent feature, making sure that no two adjacent features share the same color and that the palette is coordinated.

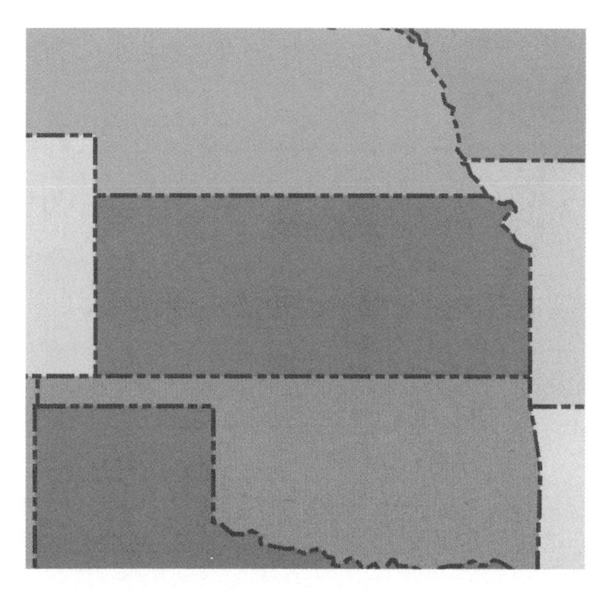

FIGURE 6.24
Along with different fill colors for each state, this map uses a dot-dash line for the state boundary lines.

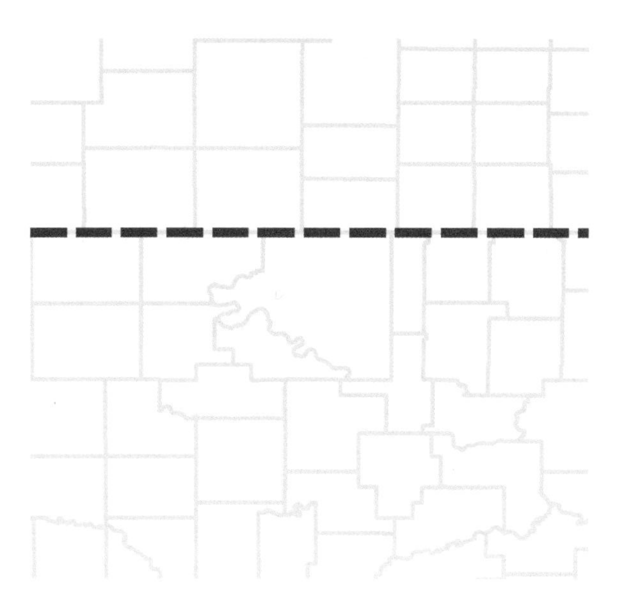

FIGURE 6.25
A nested hierarchy of political boundary types allows the viewer to easily pinpoint the major boundaries while still being able to see the minor ones. In this map, the dashed line represents a state boundary line. The dashed line could also be used on the outer boundary lines of a group of states to depict a marketing area, for example.

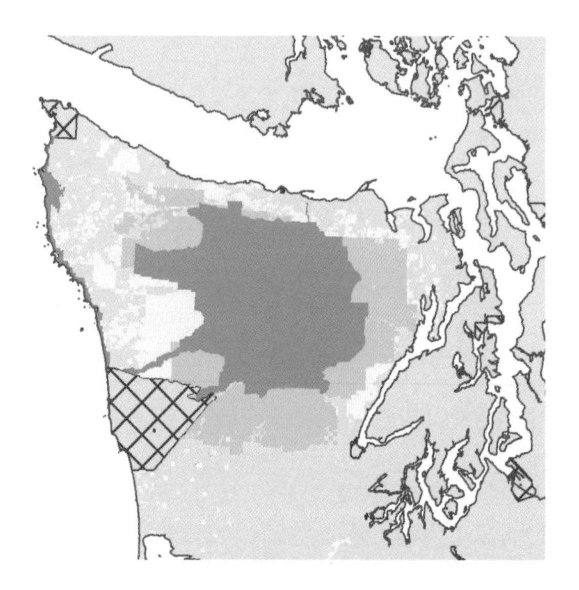

FIGURE 6.26
In this map, a lot of different political boundaries are shown. Since they are not hierarchical, they were differentiated by using fill colors for some (federal and state parks), hashing for others (tribal land), and outlines for others (state boundaries).

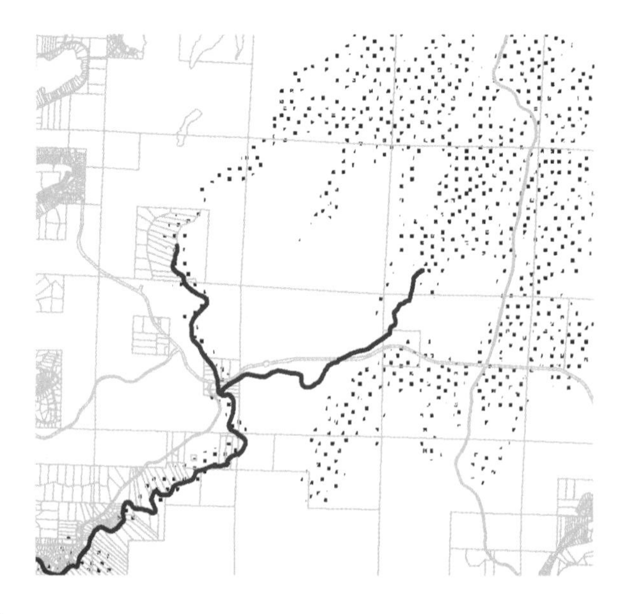

FIGURE 6.27
A random-looking dot pattern (which is really just a graphic fill) for the flood-prone zones in this map is helpful for conveying the imprecise nature of the boundaries.

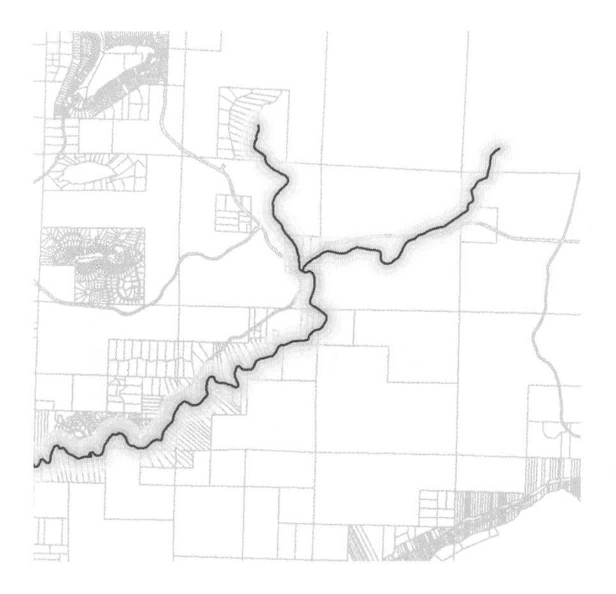

FIGURE 6.28
Stream buffers can be contentious and are not always to be taken as exact boundaries to determine on-the-ground activities. To show this 600-foot stream buffer's inherent fuzziness, a multiring buffer was created at 200-foot, 400-foot, and 600-foot intervals. The 200-foot interval is medium green, the 400-foot interval is light green, and the 600-foot interval is a lighter green. With no outlines and a transparent effect, the buffer shows its general location without appearing set in stone.

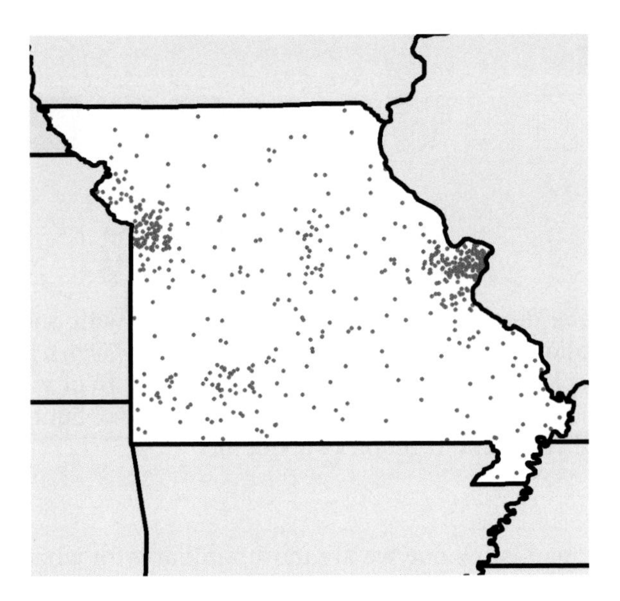

FIGURE 6.29
To get rid of any hints of boundaries, you can always transform polygon data into dot maps where dots represent a number based on a variable. In this case, the polygons aren't needed anymore.

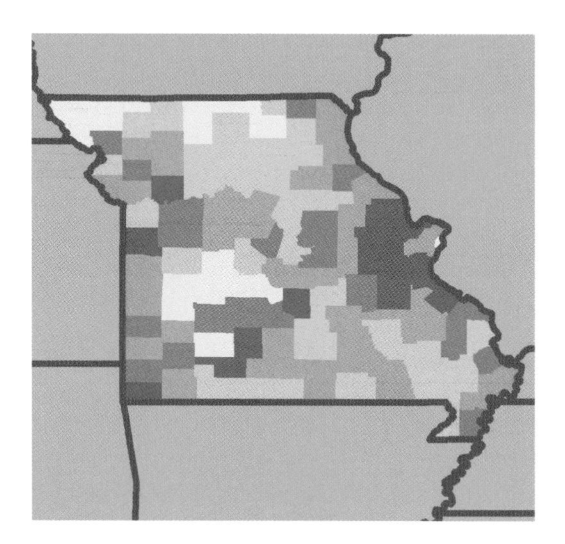

FIGURE 6.30
A simple way to deemphasize boundaries is to get rid of them, use a graduated color fill, and make sure the colors are similar enough not to create a huge difference between adjacent colors, but different enough to show the overall trend. Additionally, if the legend only specifies a start and end number, then the viewer can only infer the in-between numbers.

Elevation and Hillshade

Hillshade Colors[*]

Elevation Colors[†]

Elevation can be displayed in many different ways, with one particularly popular derivation being the hillshade (also called *relief representation*). Other options for displaying elevation include contouring with or without hypsometric tinting, spot height labels, and, of course, the 3D oblique effect (which is really 2.5D if you want to get picky about it).

Hillshade

The hillshade map is the one we are most familiar with when we think of a grayscale relief map that shows shadows on the sides of slopes to give an

[*] The color bar contains suggested colors for hillshade.
[†] The color bar contains suggested colors for elevation.

almost true-to-life feeling of height. A hillshade map is calculated from an elevation dataset using an algorithm that determines the amount of shadow to apply to each raster cell depending on the elevation value of that cell and the location of the light source—usually the northwest corner of the map. Slope and aspect values are also taken into account. The great thing about a hillshade map is that when you put it under some transparent layers, you can really get a sense of where those layers' features are in relation to the topography of an area. For example, if you had a layer classified into different types of trees and put that over a hillshade map, you can start to see that there are definite differences in tree type depending on the underlying elevation. You can also see landforms on the hillshade that act as landmarks, so that you can more easily relate the attributes of the top layers with their geographic location.[2]

Contours and hypsometric tinting: Contour lines that follow a single elevation value around a grid can be computed and displayed within the GIS and on your map. These are usually brown for elevations that are above sea level (except over glaciers, where they are blue) and blue for elevations that are below sea level. The elevations ought to be labeled for each 5 or 10 contour lines. If there are only a few contours, however, they can all be labeled. The labels are usually placed on the line with white space behind them (see Figure 6.31).

Traditionally, every fifth or tenth contour is symbolized with a thicker line or a different color than those contours that appear between. The thicker lines are the index contours and the thinner lines are the intervals. The difference in line widths and colors allows for quick slope calculations and easy visual interpretation. Contours are harder for the untrained eye to decipher than a hillshade, but they are a more accurate representation of the actual elevation than a hillshade. So your map's purpose and audience are going to determine which of these methods you wind up using.

If contours are in order for your map, you might consider coupling them with hypsometric tinting (shaded contour map or shaded isoline map), which is really just a way of saying that you could color between the contour lines to provide additional cues for the viewer who does not have experience with contours. To explain this further, picture a map with each elevation group (say every 50 meters or 100 meters) having a different color, with the highest elevations, like the tops of mountains, being white, and the lowest areas, like flat-lands at sea level, being dark green. You can continue the fun by using shades of blue for the locations that are below sea level. Hypsometric tinting is also desirable on top of hillshade layers and creates a very data-rich cartographic output.

FIGURE 6.31
A contour label placed on the line.

A lot of times when we use a hillshade map, especially if it also involves hypsometric tinting, we risk distracting the map viewer from the important aspects of the map because there is just so much color and detail all around the map. To overcome this problem, you may want to isolate the most important section of your map for hillshading. However, should you need to provide elevation data outside of the map's primary focus area to provide context, you may want to consider using a very lightly colored hypsometric tinting layer that covers the entire area, while using a darker-colored hypsometric tinting layer on the focus area.

Spot Height Labels

Labeling the important elevations on the map with their elevation values is called *spot height labeling*. Usually you would label elevations that are extremely high, like Longs Peak in Colorado, for example, or super-low like the Rift Valley. But you could also label the heights of buildings if your map is of a smoke-plume simulation, or the heights of hills and trees for a viewshed analysis, or whatever special case you might have.

3D Elevation

In GIS, the height dimension of a 3D map will most likely be defined by an elevation raster dataset. Now, there could be some exceptions to this (modeling the human eye or the hue, saturation, value [HSV] color model, for example),[3] but for the most part, when we talk 3D GIS we are talking about how our features look when draped onto the surface of the Earth. Sticking to the basics of the use of elevation in 3D maps, the first thing to do is make sure there is enough elevation change at the scale of your study area to make it worthwhile. If the land is relatively flat at your scale, you can create several maps at larger scales or extrude the elevation values to give them the appearance of being larger. The second thing is to be extremely cautious about clutter. If a regular elevation layer can distract from other data, then think about what a 3D elevation layer can do. Please see Figures 6.32–6.42 for examples of how to depict these features in certain situations.

Parcels

Colors*

* The color bar contains suggested colors for parcels.

FIGURE 6.32
Hypsometric tinting, shown here draped over a hillshade, allows the overall height category and the underlying hills and valleys to be seen.

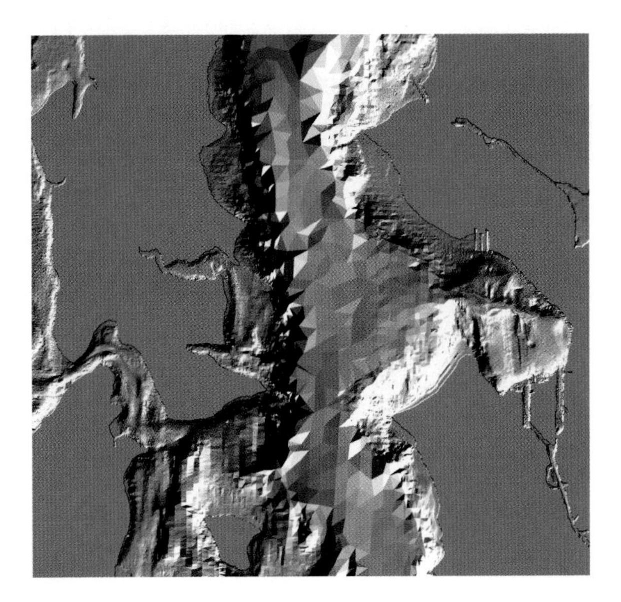

FIGURE 6.33
A hillshade for only the underwater portions of the elevation map makes for an interesting twist from the more typical map where the on-land portions are hillshaded while the underwater portions appear flat.

FIGURE 6.34
The color scheme for an elevation dataset can vary depending on the purpose of your map. In this map, an uncommon purple and green color scheme makes the map appear more art-like than life-like.

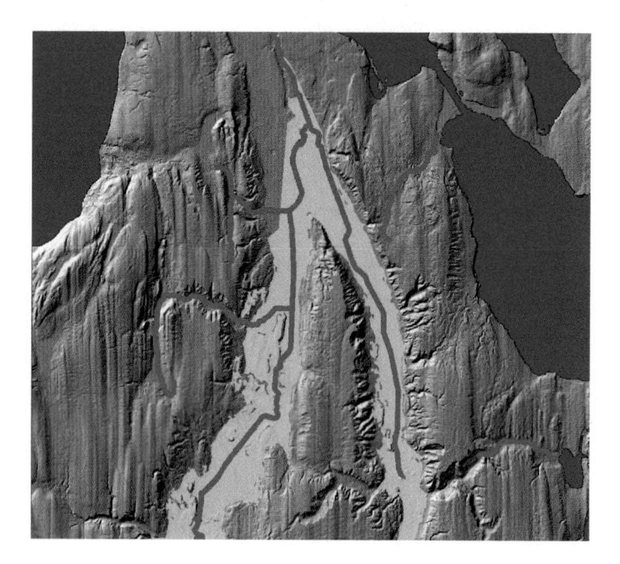

FIGURE 6.35
A hillshade is a great background layer for stream features and related hydrological features. In this map, a transparent floodplain layer is draped over the hillshade, which makes the relationship between floodplain and elevation explicit.

FIGURE 6.36
In this example, the parcels are dotted around the outskirts of town to convey the impreciseness of their boundaries, while the in-town parcels are depicted with solid lines to convey precision. The legend would indicate that the solid lines are surveyed parcels and the dotted lines are hand-drawn parcels.

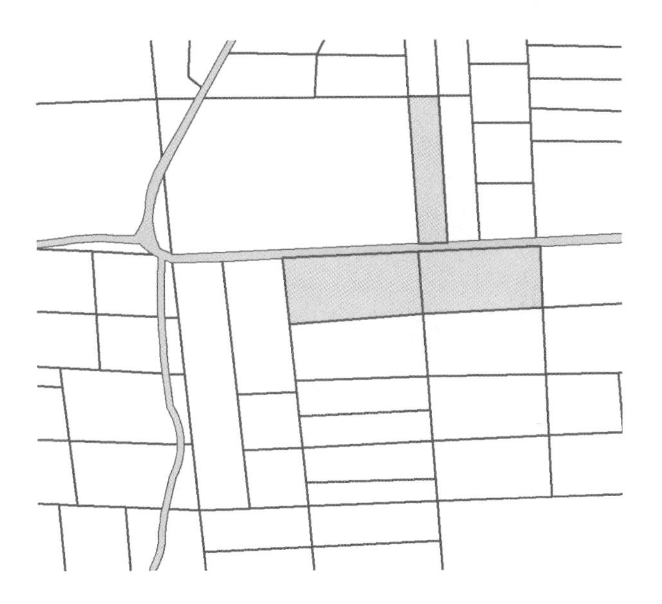

FIGURE 6.37
This large-scale map uses a highlighting color to emphasize a few parcels. The surrounding parcels are noticeable but not obtrusive.

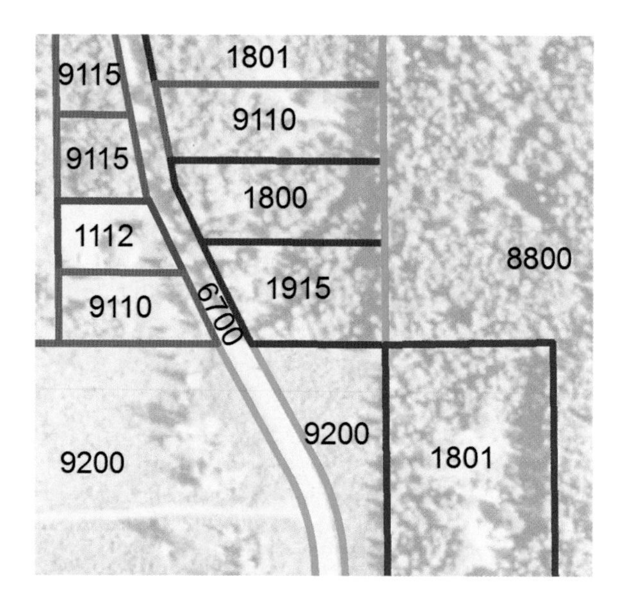

FIGURE 6.38
Just the outlines of parcels, color coded to denote regulatory zones, are shown here overlaid with a high-resolution aerial photograph. Bright parcel outline colors are needed to allow the parcels to be seen above the aerial photograph. Labels indicate land use codes.

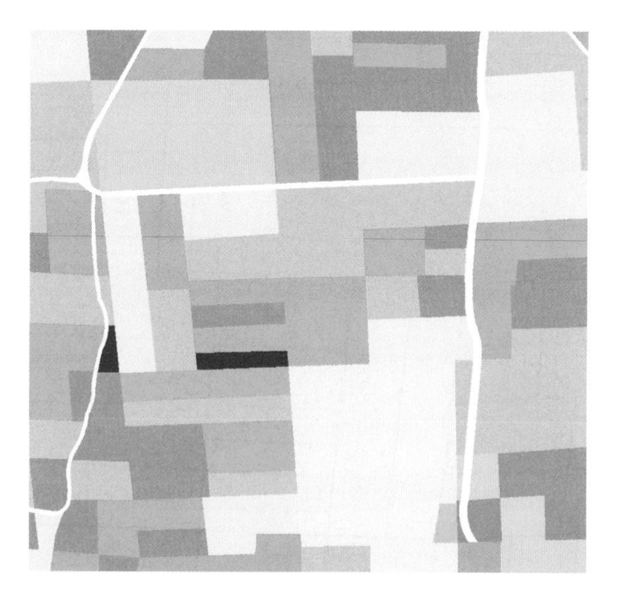

FIGURE 6.39
A patchwork of parcels can indicate general trends in land use or any other variable tied to the parcels. In this case, oranges and reds indicate residential properties by housing density, greens indicate forested parcels, and tans indicate public open space.

FIGURE 6.40
In this current map of a portion of San Francisco Bay, there are three graduated symbol sizes to provide a visualization of current velocity. More than three symbol sizes would overwhelm this map due to its size and scale. The symbols are rotated to depict direction. Simple symbols do not portray direction (such as triangles, circles, squares); you must use a symbol with a tail like an arrow or elongated triangle.

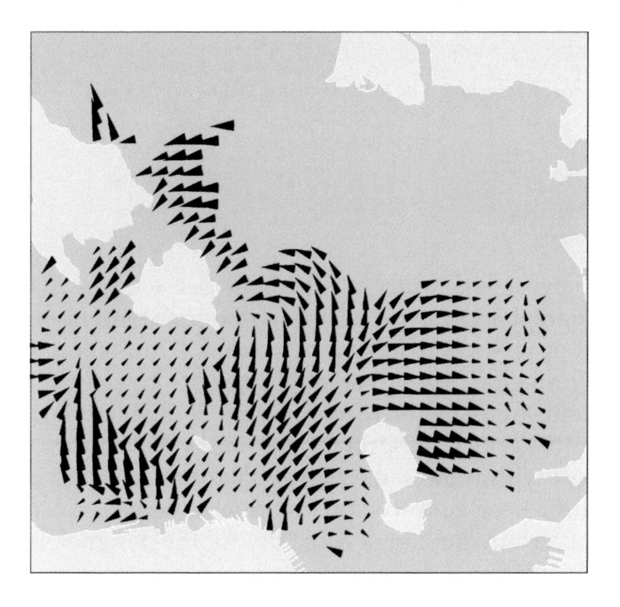

FIGURE 6.41
In this smaller-scale view of the previous map, the number of gradients was kept the same but the symbols were made significantly smaller.

FIGURE 6.42
To show the major ocean currents at a small scale, use elongated arrows as in this example, or some other symbol with a definite direction. Streaklets, symbols that look like comets, would also work nicely and have a more fluid appearance. In this map, the arrow is red to denote warm temperature and is labeled with the name of this major current.

A parcel by any other name would be just as boring: Other words that have similar meanings as parcel are

- Tax plat
- Lot line
- Property boundary
- Cadastral map

Parcels, or whatever you want to call them (see Figures 6.36–6.39), when stored in a geographic database, delineate the aerial extent of property and store all kinds of related attributes such as tax ID number, tax status, and ownership information. These parcel boundaries can be very precise due to on-the-ground surveying or very imprecise if the boundaries are instead digitized from historic, hand-drawn, plat maps. Sometimes both types of boundaries exist within the same parcel database, as in the case where in-town parcels have been surveyed on the ground while rural areas have not. This presents a certain amount of difficulty for the designer since it is likely that both types will need to be presented on the same map. A distinction could be made between the precise boundaries and imprecise boundaries by changing the boundary line from, say, a solid for precisely located parcels, to a dashed line for nonsurveyed parcels. In other cases the designer may want to merely note the limitations of the data in a disclaimer or other such section containing caveats.

Scale must be a big part of the discussion on the subject of adequately portraying parcels on a map. Parcels are really only visible as discrete entities at large scales since parcels can be as small as, say 30 meters by 30 meters, although they can be much larger. Also, your study area will most likely include a mix of small and large parcels (representing small stores downtown and large golf courses in the outskirts, for example). So what do you do if you want to show parcels on the map but have a broad study area that will effectively obscure the individual parcel boundaries? The first thing to do is figure out what it is you are trying to convey to the map viewer by means of parcel boundaries. If the intent of the parcel data is to merely show a background map to provide context for other data, then there are a few options. The first is to consider only showing the landmark-type features from a parcel map, such as the roads, parks, and water areas. The second is to show all of the parcels but only in outline form (i.e., no fill color) and with a very light gray and very thin line symbology. This second method will not work, however, with very broad study areas at the smaller scales.

Sometimes we wish to use fill colors for the parcels to differentiate parcel type, such as red for commercial and industrial parcels, green for park parcels, and brown for residential parcels. In this case, use a very unobtrusive, thin outline for the parcels, or don't use an outline at all. Thick parcel outlines coupled with loud fill colors is a sure way to Clutterville. Smaller-scale parcel maps, where the parcels are too small for any type of fill or outline, require one or more inset boxes that enlarge the denser regions of the map to allow your viewer to see the smaller parcels within those dense sections.

Above all, always ask yourself if this data could be summarized in a way that would assist the viewer in interpreting the results more swiftly. For example, the commercial and industrial, park, residential map could be displayed as a park map where the ratio of park to township is shown rather than the actual land uses. In this way, you've done the interpretation for the viewer as opposed to counting on them to interpret it themselves. Some other options are to include both the *data* map and the *derived* map on the same sheet so that the viewer understands where your analysis came from. In other cases, however, you will not be able to summarize the data, particularly if you are using the map as a means of gathering corrected information from a large group of people. For example, you may need to vet your commercial and industrial, park, and residential categories with planners and local citizens who have first-hand knowledge of the study area to make sure you have categorized things appropriately. One solution for this is to increase the page size or use multiple pages for print maps. Another solution is to create a multiscale map with parcels rendered at the higher zoom levels.

Currents

Colors*

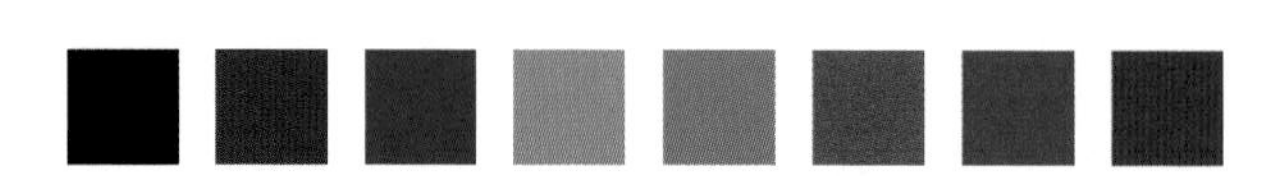

Ocean currents (and other water currents as well) are usually shown on maps to aid in naval navigation, but can also be a neat way to visualize how water currents shape the environment for other purposes. Usually, a current map will consist of text and symbols that convey the velocity and direction of currents on a given body of water (ocean or lake, primarily, but also applicable to rivers and streams if such data is available). Current direction is typically shown with arrows pointing in the dominant flow direction while velocity can be shown with symbol levels (Figure 6.43). Symbol levels can be made up of color gradients, line thickness gradients, and symbol gradients. For example, a darker blue and thicker line arrow could represent high velocity while a lighter blue and thinner line arrow could represent low velocity. Another great way to show velocity is to create a map where there are more arrows in the higher-velocity areas than in the lower-velocity areas. That sort of density-dependent symbology takes some time to symbolize properly, but is very appealing in its

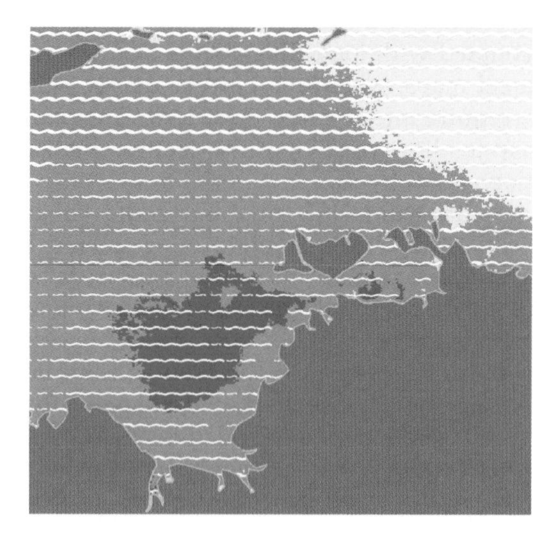

FIGURE 6.43
Ocean currents are often shown along with sea surface temperature. When using blues, purples, yellows, and reds to denote temperature, the currents need to pop out sufficiently from the dark background by using whites, yellows, or bright greens. Alternatively, the background can be lightened with a masking layer or transparency effect.

* The color bar contains suggested colors for currents.

intuitiveness to the untrained viewer. On a smaller–scale map, perhaps all you need are a few, longer arrows, showing average current direction for various known currents along a shore or in an entire ocean. Digital, interactive maps can depict the currents in a real-time, animated fashion.

Current arrows can consist of solid lines or outlines. Conventionally, the colors of the arrows are red, blue, or black. Red and blue current arrows represent warm and cool currents, respectively. Display techniques for printed maps depend largely on the scale of the map. A map of the world would show the major ocean currents and current name labels. A large-scale map showing just one beach along a shoreline, though, or one bay, could have a great number of arrows to show every little current that exists. Current velocity can be shown with symbol levels by using longer arrows (or whatever symbol you are using) for the high-velocity currents and shorter arrows for the low-velocity currents.

Wind

Colors*

Who knew there could be so much to learn about the cartography of wind? This is one particular area of mapping that must have made those old-time cartographers jump out of their seats with joy. "Aha!" I can just picture them saying, "I love putting wind currents on maps because in doing so it gives me good excuse to draw a picture of a wind god in some of the white space!" And just what wind god might they have picked? Aeolus, for one, from the tempestuous story of How Odysseus and His Crew Were Idiots (or something like that): Consider poor Odysseus who, upon having been rescued by the King of the Winds—Aeolus—and given a whole month to rest up on Aeolus's home land of Aeoli, is subsequently given some bags of the wind as a gift by the Wind King himself. After the wind gift is bestowed, Odysseus is sent nicely on his way home with a strong breeze from Aeolus, headed in the right direction. Just when Odysseus and his crew are almost home-sweet-home, the sneaky crew can't help but open those gift bags of wind that Aeolus had given them, expressly against Aeolus's instructions and Odysseus's command. And blast it, they open up winds that send them straight back to Aeolus, who wisely decides he will not help the unfortunate crew anymore (Figure 6.44).

Certainly, sailors like Odysseus must be familiar with the wind currents through which they sail. And certainly we geoprofessionals will sometimes be called upon to show these currents on maps for those sailors, but more

* The color bar contains suggested colors for wind.

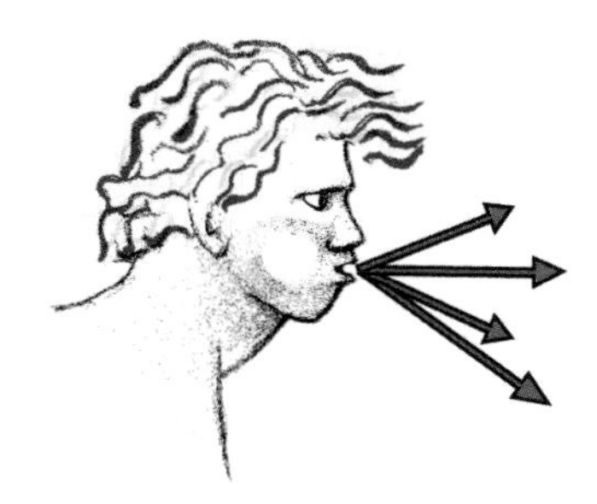

FIGURE 6.44
The red hue of these GIS-based wind arrows shows that Aeolus is blowing a lot of hot air.

frequently than not, we will be displaying wind on our maps for other sound reasons, such as displaying an analysis of wind currents versus offshore drift, or the connection between wind velocity and the tempering effects of eelgrass. Whatever the purpose for showing wind on our maps, much of the same cartographic techniques discussed in the "Currents" section are also applicable for wind data.

For one thing, wind, like currents, has direction and velocity. Both can be shown with arrows and symbol levels, respectively. That means you can use different colors and different symbol sizes for different wind speeds just as with currents. However, you can also use wind-specific symbols, called wind barbs, especially if you have wind station data tied to a particular station, or point on the map. A wind barb always points in the direction that the wind is coming from and contains barbs that indicate how fast the wind is blowing, thus showing direction and speed all in one without changing the color of the symbol. You can't read these if they are too small since being able to discern the barbs is crucial to interpreting them. Therefore they are really only useful for large-scale maps unless you make really large barbs. They can also be accompanied by labels showing a third variable.

The only thing the wind barb map in Figure 6.45 is effectively communicating is general wind direction. The wind barbs are barely discernible due to the small scale and large quantity of barbs.

So how do wind barbs work? For starters, they consist of lines that are oriented in the same direction as the wind, along with short lines coming off the ends of the direction lines, which indicate velocity. Shorter barbs indicate 5 knots and the longer ones indicate 10 knots. There can be some variation in where these velocity indicators are placed when there is more than one on the barb. For example, a 15-knot wind would be represented with two lines coming off of the main directional line, one short, and one longer. The short line could be placed at the end of the directional line with the longer one underneath, or vice versa. Another variation you might see is a dot, or small circle, at the end of the direction line. Sometimes the dot indicates a third variable such as cloud cover. A simple wind-barb example is shown in Figure 6.46, showing wind from the northeast.

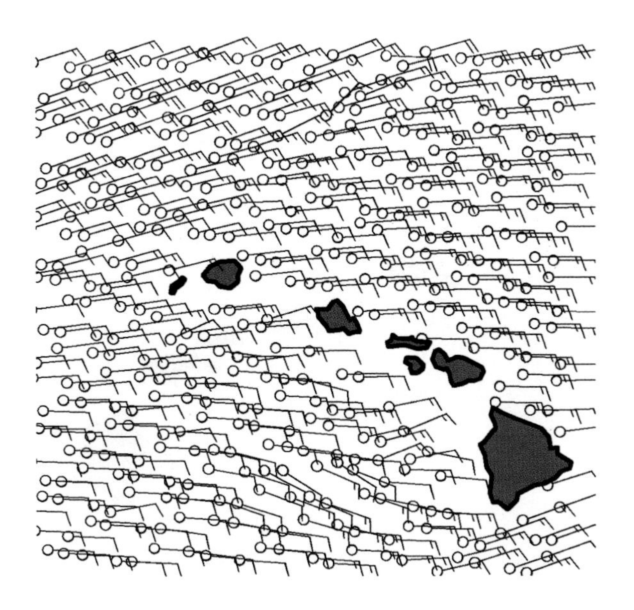

FIGURE 6.45
Too many wind barbs at this scale make for a cluttered map. To remedy the problem, some sort of generalization is in order here.

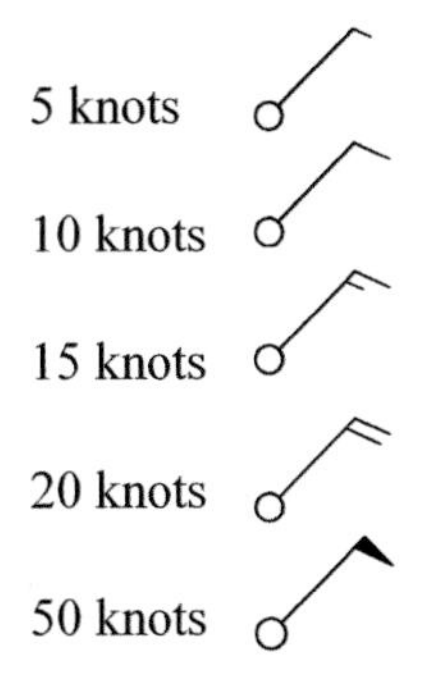

FIGURE 6.46
Wind barbs show the direction of the wind (from the northeast, here) and the speed of the wind. The angle of the main line is rotated relative to the stationary dot depending on the wind direction. The flags change in number, style, and position depending on the wind speed.

What else can be used to visualize wind on a map aside from arrows and barbs? You might sometime come across a very interesting method of showing wind using isolines (see Figures 6.47–6.53). Essentially, wind direction and velocity can be depicted with isolines connecting points of equal value. The direction isolines are called *isogons* and the velocity isolines are called *isotachs*. You might also see the term *isogon*, which actually refers to any line that depicts the angle of something, not just wind. When a map has many isolines, a hierarchy of symbols can be used to more clearly delineate the information by means

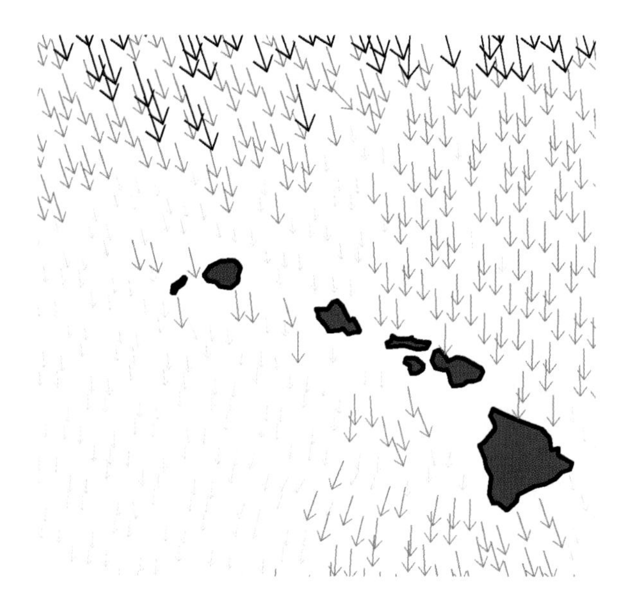

FIGURE 6.47
In this map, the wind direction is shown by means of a simple arrow rotated according to a direction field in the database. The wind speed is shown through two simultaneous cartographic effects: darker gray and longer arrows represent a fast wind, while lighter gray and shorter arrows represent a slow wind.

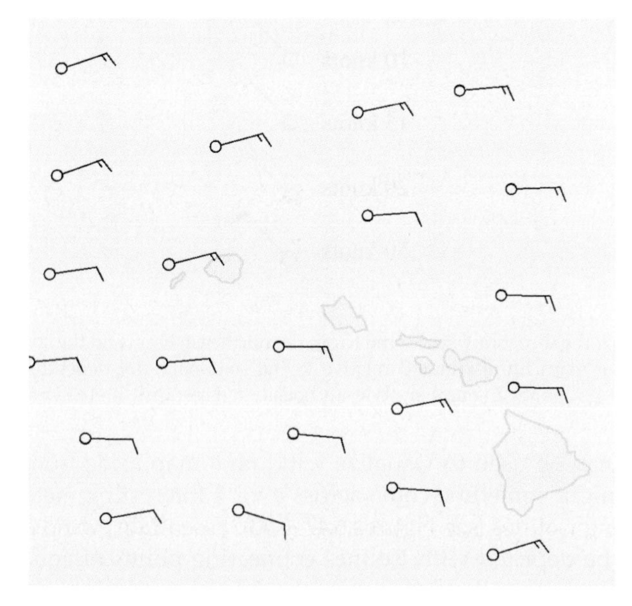

FIGURE 6.48
This wind data has been resampled to reduce visual clutter. The general wind directions and wind speeds are readily apparent.

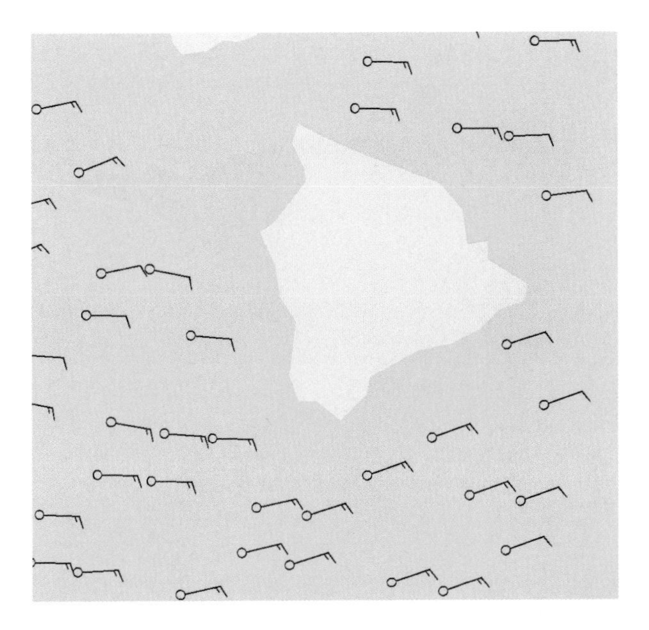

FIGURE 6.49
This wind data has not been resampled, but the map extent has been enlarged so that all of the barbs are visible and readable.

FIGURE 6.50
These temperature isolines represent varying degrees of Fahrenheit throughout Minnesota. If a general overview is in order, then this map does the trick, especially when paired with a key to the colors.

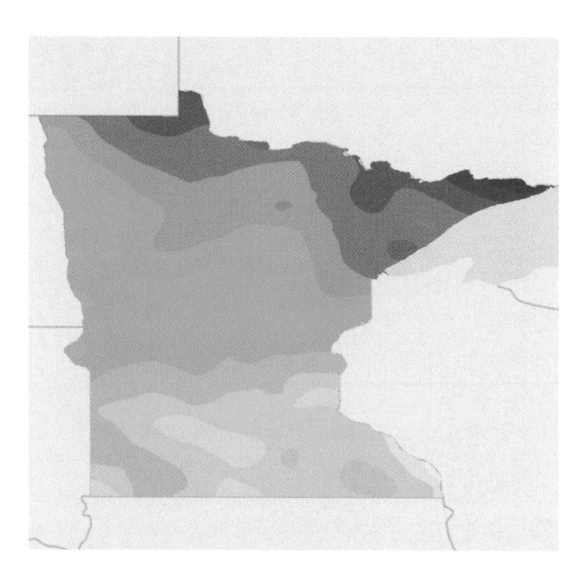

FIGURE 6.51
The areas between the isolines are shaded in this map to produce more of a continuous surface effect. This map looks great, but potentially misleads a map viewer into thinking that temperature was measured continuously across the space, whereas it was only measured at points along the lines between the color changes. Consider your audience when deciding if it is an appropriate trade-off.

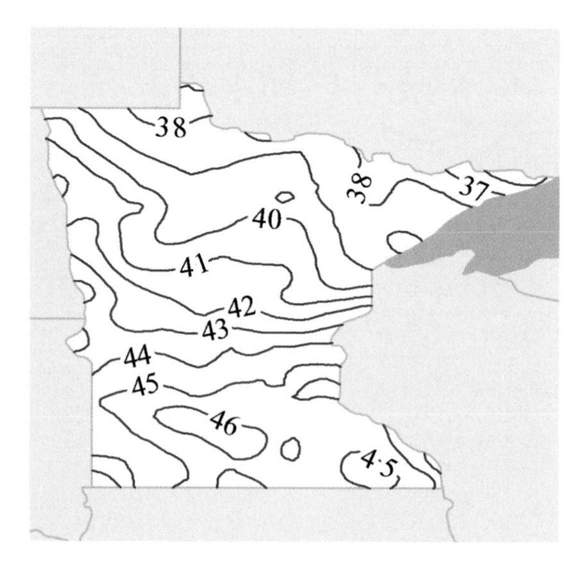

FIGURE 6.52
This black-and-white temperature isoline map has labels that show what the temperature is along that line. Although some GIS software can place labels like this automatically, you may need to use halo-text for the labels (especially if using graphics software), or use a symbol with text that has a white background (such as a road sign symbol without the outline).

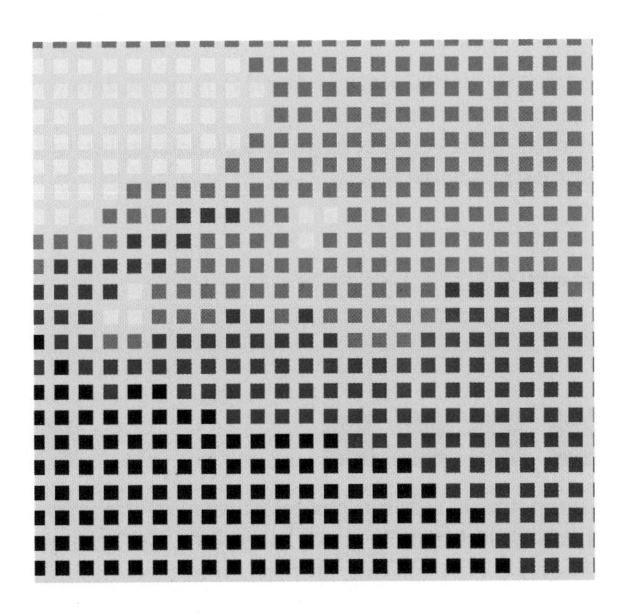

FIGURE 6.53
While not quite the same continuous surface effect as a polygon fill would produce, this map of temperature points achieves a similar effect. It also clearly conveys the message that the data is taken at discrete station points in the study area.

of a thicker line for every fifth or tenth isoline and labeled with the appropriate value. For smaller-scale maps you can show just the major intervals—all of which should be labeled—in order to avoid cluttering the map. These lines are typically black or blue, though gray might be a nice choice as well.

Hypsometric tinting can be used for these maps, just as with any isoline maps, by way of using different colors for each interstitial space between the isolines. If you use this technique, you do not necessarily need labels on the isolines because you will furnish the layout with a legend that relates the colors to the associated variable.

OTHER WEATHER FEATURES

Wind, of course, is not the only weather variable that is frequently mapped in GIS. Other features like frontal locations, pressure, and precipitation to name a few, are common inputs to GIS. The same principles that apply to wind would apply to these other features:

- Reduce clutter by resampling or zooming in when needed.
- Make sure you check out industry standard symbols before attempting to use your own.
- Simplify the symbols if needed.

- Investigate alternative ways of displaying point data (by isolines or hypsometric tinting, for example).
- Experiment with color symbol levels and styles to illustrate intensity of a variable.

Temperature

Colors*

Traditional air temperature mapping palettes are interesting because of the blue to red color scheme that is usually used. Essentially, the level of heat on a temperature map is denoted by making all cool zones blue and all hot zones red. With the addition of in-between temperature zones, the color scheme winds up being something like this: blue, yellow, orange, red. However, if you think about it, this color scheme is directly the opposite of the colors that a flame takes on at varying temperatures. A flame, such as a candle flame, can take on the colors blue, yellow, orange, or red, but the blue color indicates the hottest part of a flame while the red color indicates the coolest part of the flame (incandescence). Similarly, a look at the field of astronomy does not help either as we wind up with the same issue as with the candle flame: the hot stars are actually blue and the cool stars are actually red. So where did this standard map temperature color scheme come from? In contrast to the color properties of natural phenomena, the color properties of emotions are quite different. The emotional connotation for red, yellow, and orange is hot while the emotional connotation for blue and purple is cool. So, while it would seem to be a better fit for mapping natural phenomena in their "natural" hue, we continue to use the traditional color scheme based on emotional connotations for temperature maps.

Temperature can be symbolized in a variety of ways including

- Points where temperature readings were measured
- Lines connecting measurement points of equal value: isotherms
- Polygons with hypsometric tinting: the area between isotherms

All of these geometry types could be shown in color or in black and white. If color is used, it would be wise to use the traditional blue to red color scheme just discussed, tying the color to the temperature of the individual point, line, or polygon. Grayscale palettes can be used to denote temperature as

* The color bar contains suggested colors for temperature.

well, with the darker grays denoting higher temperatures than the lighter grays. Grayscale palettes are fine for use with any of the three geometry types. Black-and-white maps can easily be constructed with temperature isotherm (line) data since the isotherms themselves can be black and simply be labeled with the appropriate temperature. While the examples in this section focus on air temperature features, maps of sea surface temperature, or any water temperatures for that matter, are mapped in exactly the same way.

Land Use and Land Cover

Colors*

Land use land cover (LULC) data consists of grids or polygons representing type and extent of terrain for a portion of the Earth's surface. There is definitely some confusion concerning what exactly the difference is between the terms *land cover* and *land use*. Land *cover* technically consists of anything that covers the land, both natural and human-made, with perhaps the exception of open water and bare rock (although, these too are sometimes included in the definition). Land *use* is supposed to be solely the activities that humans use the land for like hunting, fishing, and farming, but arguably also consist of categories such as urban, industrial, and residential that aren't exactly activities but are definitely human-centric uses of the land. Because of the slightly unclear nature of the terms, many simply concatenate them into one overarching term that groups the two kinds of information together.

Some Land Use Land Cover Categories

- Urban
- Residential
- Commercial
- Cropland
- Forest
- Ocean
- Tundra
- Ice

LULC data is derived primarily from satellite or aerial sources. Maps of LULC can drill down to the details of a landscape by providing information on the makeup of individual plant communities, for example, or take a wider view and categorize areas into broad swaths such as water, forest, urban,

* The color bar contains suggested colors for land use and land cover.

and grassland. Often, specific activities (land uses) that could potentially be included like hunting, are not, due to the lack of ancillary data needed to define the boundaries of those activities (such as administrative boundaries) as well as their overlap with other categories like forests and grasslands. It is wise to keep these caveats in mind and perhaps call them out in the text portion of your map if the audience is unfamiliar with such data.

You can think of LULC maps as being similar in visual style to any class-type map, except that in many cases they utilize standard classifications with standard color schemes. Should you not have data that uses a standard color scheme, then the convention for deriving your own colors is to keep the colors as logical and close to reality as possible, whereas with other class map types you might just assign colors based on how nice they look together. For your LULC data, forests could be varying shades of green depending on the forest type, while glaciers would be white, and bare ground would be brown, for example. To repeat, there is a higher degree of matching map colors to real colors that goes on with these maps than with other class-type maps.

As far as standardized class schemas and colors go, one of the most familiar schemas for vegetation classification of remote sensing data is the Anderson classification system, developed by a whole crew of agencies including the USGS, National Aeronautics and Space Administration, Natural Resource Conservation Service, Association of American Geographers, and the International Geographical Union.[4] With so many agencies working on this thing, it is a wonder it came out as splendidly as it did!

In this hierarchical schema, landscape features are grouped into varying levels of detail. In the Anderson nomenclature, each hierarchy step is called a *level*. For example, level II Anderson, unit 10, is urban land. This unit is broken down into eight constituent parts: low-, medium-, and high-density residential; commercial, industrial, and institutional; extractive; and open urban land. The fact that these classifications remain constant over a wide variety of analyses strengthens their usefulness in, for example, trend mapping over time.

If your data is in an Anderson classification, then you are in luck as far as picking colors is concerned, because the USGS has done the job of picking the colors you should use for you, at least for the level I categories. The official colors are reported in Munsell color format, but I have converted them to GIS-friendly RGB triplets, as shown in Figure 6.54.

Another LULC dataset, the National Land Cover Database (NLCD), has standard categories and colors as well. While there are four levels of Anderson classification comprising more than 100 LULC categories, there are only 29 categories in the 2001 NLCD dataset and 16 categories in the 2006 NLCD dataset. In the Anderson schema, levels I and II are land cover–type categories and levels III and IV are land use–type categories. The NLCD—a derivative of the Anderson classification—does not use land use categories, ostensibly because it would be too difficult to obtain land use information on a nationwide scale at 30-meter intervals. The NLCD, therefore, has fewer levels and categories (see Figure 6.55).

Land Cover Category Name	No.	RGB Color
Urban or Built-up Land	1 ■	235 109 105
Agricultural Land	2 ■	206 166 138
Rangeland	3	255 224 174
Forest Land	4	166 213 158
Water	5 ■	117 181 220
Wetland	6	188 219 232
Barren	7	200 200 200
Tundra	8	189 221 209
Perennial Snow or Ice	9 □	255 255 255

FIGURE 6.54
These are the categories that make up the level I Anderson classification along with the colors originally assigned by the US Geological Survey. The US Geological Survey colors are in the Munsell color model. Since GIS software does not typically utilize that color system, I have converted the colors into RGB triplets.[11] Conversion between color systems is never exact, so this is an approximation of the Munsell colors.

Land Cover Category Name	No.	RGB Color	Land Cover Category Name	No.	RGB Color
Water	11 ■	102 140 190	Shrubland	51	220 202 143
Perennial ice, snow	12 □	255 255 255	Orchards, vineyards, other	61 ■	187 174 118
Low intensity residential	21	253 229 228	Grasslands, herbaceous	71	253 233 170
High intensity residential	22 ■	247 178 159	Pasture, hay	81	252 246 93
Commercial, industrial, transportation	23 ■	231 86 78	Row crops	82 ■	202 145 71
Bare rock, sand, clay	31	210 205 192	Small grains	83 ■	121 108 75
Quarries, strip mines, gravel pits	32 ■	175 175 177	Fallow	84	244 238 203
Transitional	33 ■	83 62 118	Urban, recreational grasses	85 ■	240 156 54
Deciduous forest	41	134 200 127	Woody wetlands	91	201 230 249
Evergreen forest	42 ■	56 129 78	Emergent herbaceous wetlands	92	144 192 217
Mixed forest	43	212 231 177			

FIGURE 6.55
These are the 21 categories that comprise the 1992 NLCD along with their assigned colors in both visual and RGB triplet format as defined by the US Geological Survey. These colors can be used for the 2001 and 2006 NLCD as well.

PIXELATED MAPS

If the data is looking grainy on-screen, don't forget that the algorithmic generalization of the data upon import, such as pyramid building, can cause this effect. If that is the case, you will find that the data is smoothed upon export in most cases. Another technique to smooth pixelated data is to use a Gaussian estimation such as the Gaussian blur option in Adobe Photoshop.

The depiction of these land cover categories gets too complicated if you add too many of them. You should use the first level of classification for small-scale maps that are being used for visualization purposes (as opposed to analysis). If you have to, modify the second level into something more manageable by combining categories and thereby generalizing the data.[5]

Another consideration in the presentation of LULC in map form is to be sure to include enough metadata to allow the map viewer to really understand what the categories mean. Categories are not always as intuitive as we might think. For example, in Britain, a standard forest category may not actually have any trees in it, while in Scandinavia an area with slow-growing trees may not be included in a forest category at all.[6] As another example, if you have a seral-stage breakdown into such categories as mature canopy and immature canopy, you will want to define what these mean on the map in terms of tree diameter or tree age, because any scientist who is using your map for analysis later will want to know. It is prudent to include a citation to the appropriate LULC technical report as well so that viewers can look up the specifics like error calculations and processing algorithms that may affect their interpretation of your map. (See Figures 6.56 through 6.60 for examples of how to depict this feature in certain situations.)

Trails

Colors*

Trail features are longitudinal areas or lines that people traverse on the ground on foot, bicycle, horse, or any means other than by car. They can be paved or unpaved, in a wilderness area or in the heart of a city. On a GIS map, they are typically depicted as dotted lines in order to distinguish them from

* The color bar contains suggested colors for trails.

FIGURE 6.56
This is the 1992 NLCD for South Carolina. It is displayed here using the standard US Geological Survey NLCD colors (see Figure 6.68).

FIGURE 6.57
This is the NLCD for Colorado using a black-light filter on the suggested color scheme in a graphics program. Graphics color filters provide easy ways to explore color changes without assigning individual colors yourself.

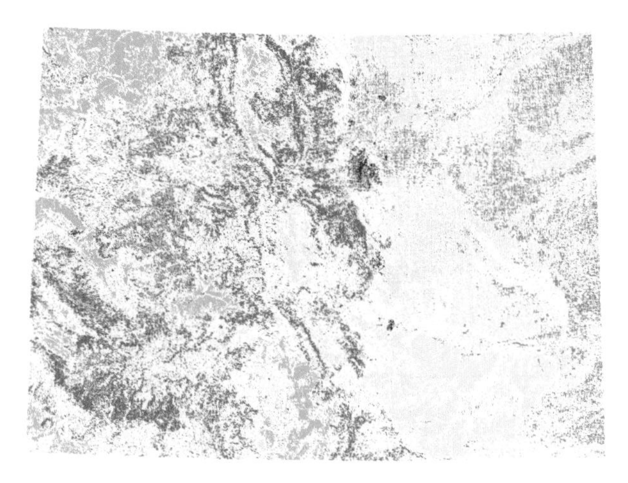

FIGURE 6.58
This is the NLCD for Colorado using the suggested color scheme, which is a natural color palette. The benefit of this palette is its familiarity and ease of deciphering. The benefit of the more modern palette in Figure 6.70 is its high contrast, providing for rapid visual scanning of a few of the NLCD classes.

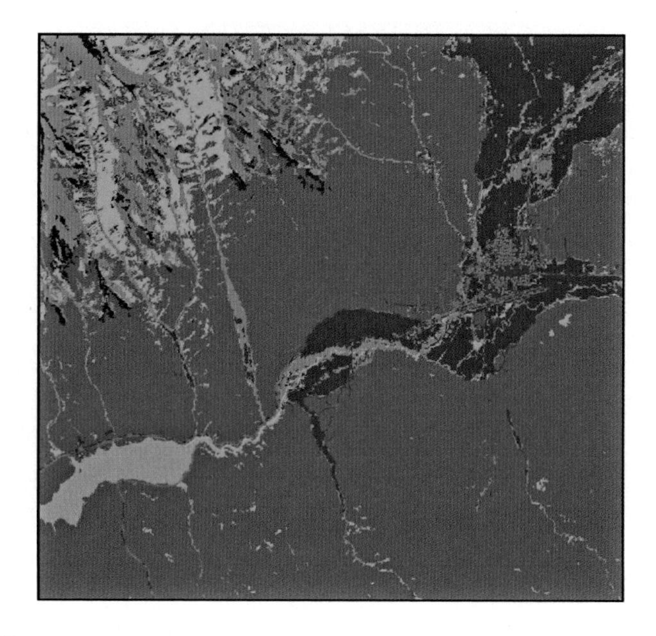

FIGURE 6.59
This is the NLCD Colorado data zoomed in to a 1:500,000 scale using the modified color scheme.

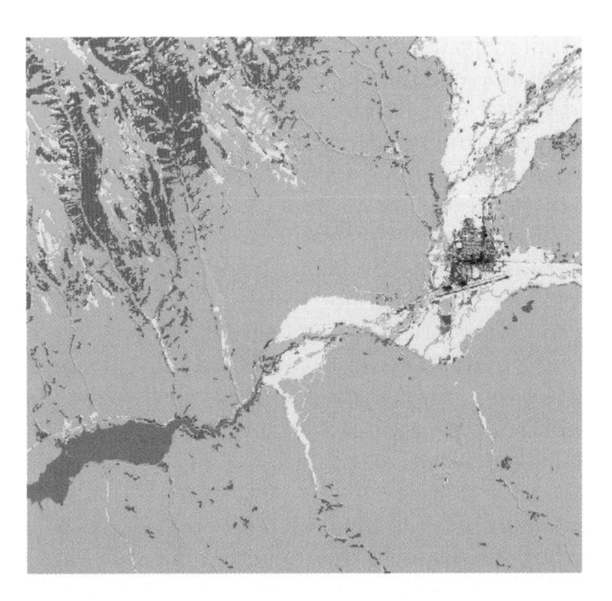

FIGURE 6.60
This is the NLCD Colorado data zoomed in to a 1:500,000 scale using the suggested color scheme.

paved roads, which are, in turn, usually solid lines. However, a trail map in a nondeveloped area such as a national park may depict the trails with solid lines since, in this case, there are no other solid-lined features to display.

Trails are labeled similarly to rivers and streams, with the labels placed above the trail lines and following the trail curves. If roads are also shown on the map, the trail's label will be differentiated from the road labels by ensuring the color of the text is different from the color of the road label text (and, in most cases, matching the trail color). Or, if color differentiation is not possible, then a different font can suffice. These trail names are usually found again somewhere in the map's margin elements, where they are listed along with details about each trail. Alternatively, numbers are used in lieu of trail names and serve the same function. If numbers are used as labels, then it is prudent to place the number at the beginning node and along the trail feature at sufficient intervals to enable rapid trail identification.

Varying shades of black and brown are usually used for trails, with brown being the obvious choice since many trails are indeed brown on the ground. Red and dusty orange, though, are also good choices, especially for maps with many features to differentiate or when the trail needs to be especially highlighted. Trail features are usually accompanied by ancillary features such as trailhead points, parking areas, and camp site points so that the map can be used for navigation. In such maps, photographs of the landscape surrounding a trail at key points, like vistas, lend themselves nicely to the layout margins of a trail map or to the click events of interactive maps. In terms of line thickness, trail maps used for navigation can depict trail lines with thick and bold lines, just as a road map would display its roads. Or, a trail map with both roads and trails can differentiate the two with a thinner line style for the trails to emphasize the smaller width of a trail in relation to roads. These options depend on the intended focus of the map and the amount of real-world likeness that is needed or desired. Please see Figures 6.61 through 6.68 for examples of how to depict this feature in certain situations.

One of the very interesting problems associated with features like trails that are very narrow, long, and not widely dispersed, is how to highlight them without throwing off the map's visual balance. As a counterpoint to this, roads, which are also very narrow and long features, are usually well dispersed across a map surface, enough so that the visual impact is more consistent across the page. A trail map, though, might consist of only two or three bike trails that pass through a city, for example. The ratio of city to trail area for a case like this is very large. You might want to show the entire city area in order to give the map viewer a good geographic reference and to show the trails in their entirety, but still want the trail to be the main focus. If all of the trails are concentrated in one portion of the city, then the visual balance will be off. One approach to minimize that problem is to use a very consistent visual feature as the background layer, such as a lightened aerial photograph or lightened parcel layer, with the trails symbolized boldly via a highly contrasting color. The consistent background will

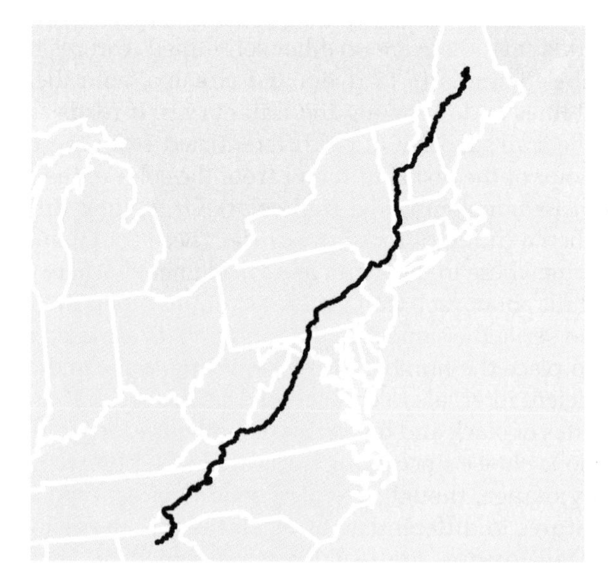

FIGURE 6.61
In this overview of the Appalachian Trail, the state outlines are shown in white so that they will not compete with the trail in visual weight. At this scale and size, no other features (such as cities) are shown, as they, too, would have distracted attention away from the trail.

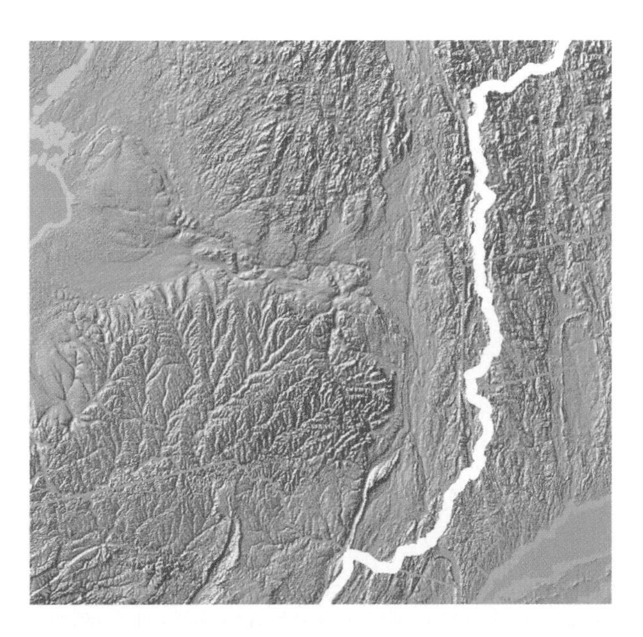

FIGURE 6.62
The intent of this map was to show the portion of the Appalachian Trail that runs through New York State, in relation to the state as a whole. Since the trail is not centered on the map, a hillshade was used as background data to provide some visual cohesiveness. Originally, the state boundaries were in white and the trail was in orange. However, this did not provide enough emphasis on the trail, so the colors were switched, with the effect that the white trail is highlighted and the state boundaries recede into the background.

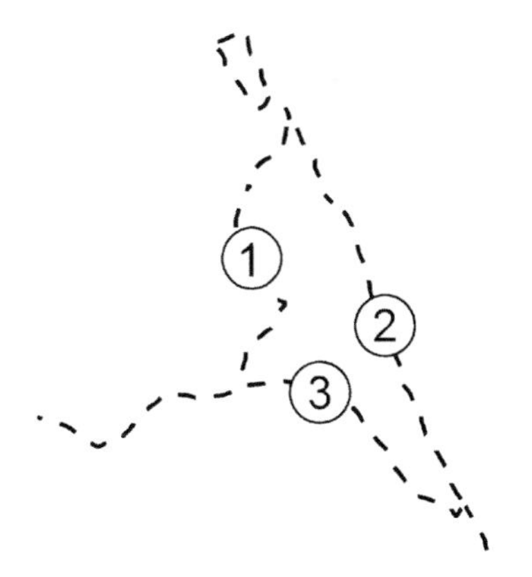

FIGURE 6.63
A common way to illustrate a trail system is with dashed brown lines. When the scale is too small to adequately present the trail name labels, a number can be used to reference the trail name elsewhere on the map. In this case, the numbers are circled to provide added visual clarity.

FIGURE 6.64
This trail map incorporates the nearby road feature and some campsite points. The solid gray line for the road and the dashed brown line for the trail are intuitive symbols, as are the orange triangles for the campsites. This also illustrates the importance of ordering features correctly: campsites on top of trails, which are, in turn, on top of the roads. This allows everything to be seen without any loss of meaning.

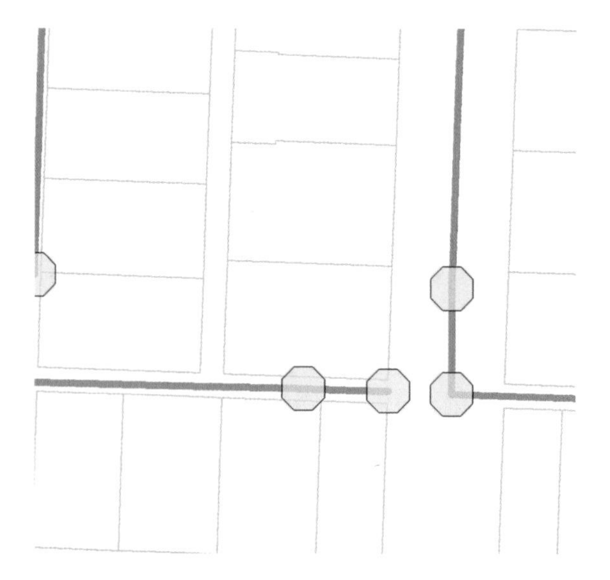

FIGURE 6.65
Transparent manhole symbols make the sewer main connections visible beneath them. The manholes are quite a bit larger in proportion to the rest of the map so that they are visible.

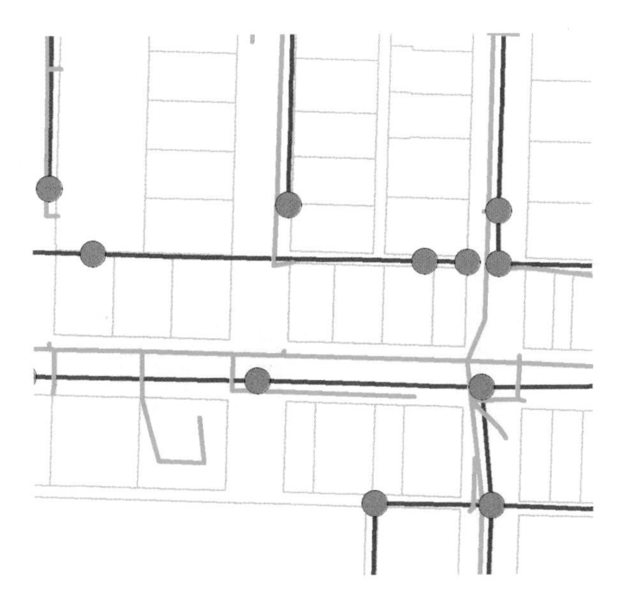

FIGURE 6.66
In this example, the storm drains and sewer drains are different colors while the symbol is the same. The manholes are not transparent and are smaller than in the previous example to provide equal focus to the storm and sewer drains.

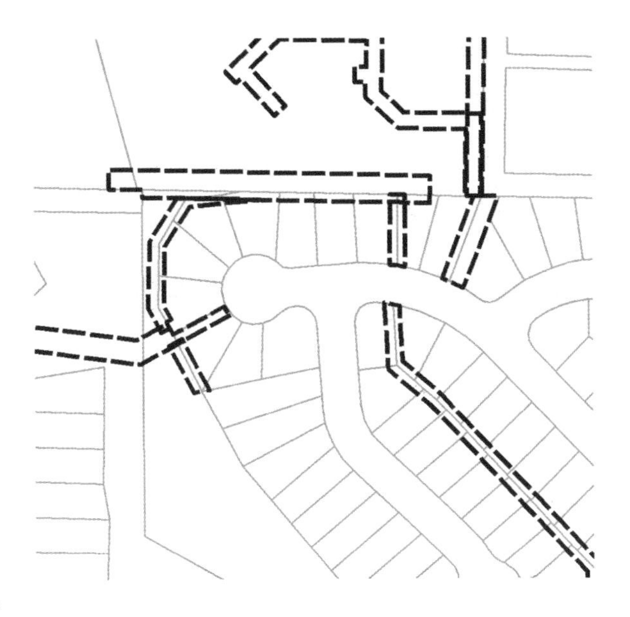

FIGURE 6.67
Easements can be shown in a dashed-line pattern to differentiate them from parcel lines. In this case, the easements were converted from polygon format to polyline format so that the dashed-line pattern would remain consistent (not overlapping).

FIGURE 6.68
The sewer zones in this map are outlined in thick black and filled with bold colors. These two techniques serve to emphasize the sewer zones, since they are the main focus of the map, while deemphasizing the supporting detail like the parcel polygons.

balance the anomalous trail features while still allowing the trails to take center stage. This kind of technique applies to any type of feature that risks getting lost due to its small size and sparseness.

Utilities

Colors*

Utilities data, such as gas and electric, water and wastewater, and telecommunications, are routinely used in a variety of GIS applications. Because utilities are such important components of our daily lives, the accurate portrayal on maps of the infrastructure and analyses associated with these features is of utmost importance. Regular maintenance, emergency repairs, personnel allocations, and potential sites for expansion are all examples of maps that use utilities as their basic building blocks. Symbols for all aspects of utility features management are highly standardized and codified. Always check for standard symbol sets and palettes prior to map production.

* The color bar contains suggested colors for utilities.

Some of the challenges include adequately representing vast amounts of data over a small amount of space, such as multiple overlapping power lines between two poles or many pipelines underneath a single city street. In fact, features like power lines are so thin in real life that depicting them on a map can pose problems. Usually we have it the other way around with maps: the features we are representing graphically are actually much larger in proportion to the landscape than they appear on the map. However, with power lines, gas lines, wastewater pipes, and the like, we will need to use thicker lines rather than the ultrathin lines that would result if everything were in proper proportion to reality. Similarly, dots that show the locations of well heads, storm drainage outfalls, or manhole covers will be shown as larger-than-life in order to make them visible on even the largest scale maps. Even on large-scale maps, however, you may run into the problem where the features wind up overlapping each other if you make the symbol sizes too big. If a happy medium cannot be found between making the symbols large enough to see and making them small enough so that they don't overlap, you may need a larger scale or transparent symbology; or you may have to split the data into multiple maps based on a feature attribute. Alternatively, consider a multiscale, interactive map, that depicts the features as groups at the smaller scales and as separate entities as the user zooms in to the highest zoom levels.

The use of color to differentiate types of features that would otherwise look the same is extensively used for utilities data. For example, fire hydrants might be colored yellow, red, green, and blue to differentiate between the flow volume capabilities of different hydrants. Linear features like pipes could be colored differently depending on direction of flow or outfall. Typically, linear features like roads, rivers, and streams are depicted on maps without their associated nodes. However, the nodes for utility line features are often explicitly drawn on top of the lines to emphasize pipe connections, poles, or manholes that occur at those nodes. Many utility maps are so complex that a simple road background (or nothing at all) is used for context. Topographic maps, orthophotos, or parcels might also be used as basemaps depending on the map's purpose and scale.

Field Mapping

When it comes to utilities, the traditional paper map is not used in the field as much as it used to be. Mobile devices are much more able to handle the complex data inputs and the multiscale needs of the utility field crew while allowing real-time data updates to be displayed for other crew members and even the public as work is occurring. Maps on these mobile devices may need bright colors and large symbols for the crew to adequately read them in bright sunlight and in darkened vehicles.

Utilities maps can get so complex with all the lines, dots, crazy symbols, colors, and whatnot, that sometimes I wonder why and how people manage to stick even more complicated data like aerial photographs or contours underneath it all. I believe that the use of those basemaps underneath such highly complex and detailed data do nothing to enhance the readability of the map. As always, you must be conscientious of your map audience and your map's purpose. If you want to show the city council where new fire hydrants will be placed, then do you really need 0.5-meter aerial photographs underneath the hydrant dots? On the flip side, that data could be useful if, for example, a group of experts will be poring over your maps in a meeting to help decide where to locate a new utility line. Remember to always have a reason for doing things. Do not place high-resolution aerial photographs on your map *just because.* Yes, people can handle the information, but will it be as effective as possible? The best way to know is to actually try the map with and without all the layers. Render it in several iterations with different layers clicked on and off before deciding.

Impervious Surface

Colors*

Impervious surface data is a popular dataset for analysts to use in all kinds of computations from measuring growth over time, to determining hot spots, to identifying illegal building activities. Showing these datasets on a map is not a difficult task, although in a highly populated area, you may find that impervious surface pixels or polygons tend to overwhelm the map if there are other details that you wish to highlight as well.

Some impervious data is organized as Boolean pixels: a pixel either is or is not impervious surface. Other impervious data represents a range of imperviousness within the pixel, usually in categories such as 0% to 10%, 10% to 50%, and 50% to 100%. Boolean data is displayed as solid-color pixels, usually in a dark gray, black, or red. Range data is displayed with gray or single-hue color ramps. Ancillary data will help the viewer to understand the data better if the viewer is not immediately familiar with impervious surface data. For example, having a roads layer or parcel layer underneath the impervious pixels allows the viewer to see that the impervious pixels overlap with the roads and rooftops. If the map is at a small scale, then country boundaries and city points will help the viewer see that impervious surfaces are concentrated in cities and highly developed countries.

When displaying a series of impervious surface maps to illustrate change over time, my recommendation is to get rid of all superfluous details. Think very carefully about each and every element on the map element to decide if it can be taken off. All the scales will match, of course, so the scale bar can be

* The color bar contains suggested colors for impervious surfaces.

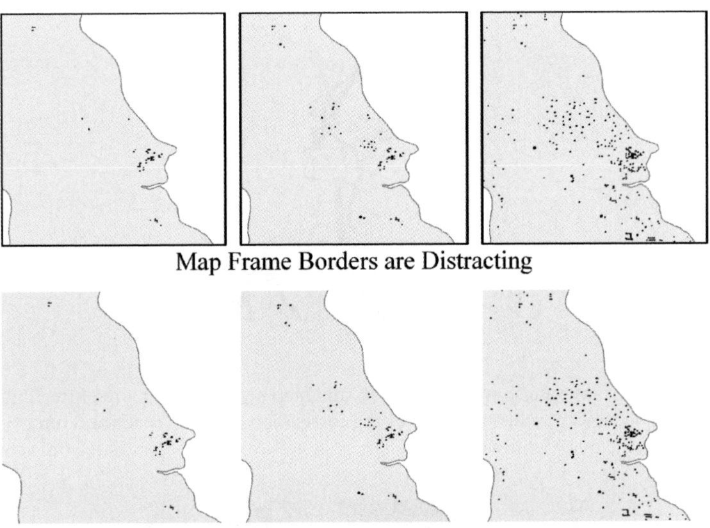

FIGURE 6.69
Impervious surface data is often presented as a series of maps over time. When constructing a map such as this, be sure to evaluate the need for map frames. There usually is no compelling reason to use them and they only serve to distract the viewer's attention from the main focus of the map.

placed in the margins of the map page. The north arrow is another element that will be the same regardless, so it can be relegated to the margins as well. Most importantly, see if you can get rid of the map frame. Your map viewer will then focus on the message—that impervious surfaces are increasing in certain areas over time, for example—rather than the graphic elements that only serve to clutter up the page (see Figure 6.69). Please see Figure 6.70 through 6.73 for examples of how to depict this feature in certain situations.

Basins

Colors*

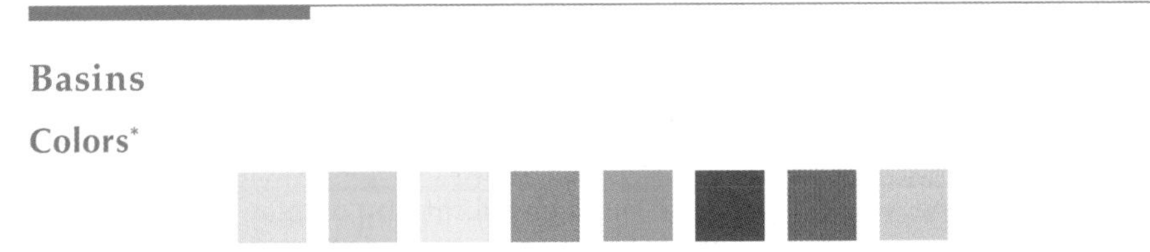

Basins are areas that are delineated via topography and define areas in which water flows downward toward a common point. Basins are also referred to as watersheds and catchments. They are a unique feature type that is often shown on natural resources–related maps. As with all features, the basin

* The color bar contains suggested colors for basins.

FIGURE 6.70
The impervious surface data shown in black on this map are layered on top of the highway and road features. The viewer automatically sees the correlation between roads and impervious surface. In this case, the map defines the term *impervious surface* much better than words could have.

FIGURE 6.71
The NLCD 2001 contains impervious surface data in percentage form. This makes it easy to use a color ramp to represent varying degrees of imperviousness. In this case white, gray, red, and dark purple form the color ramp for this densely urbanized area.

features might be the main focus of the map or just supporting information. When you create new basins with your GIS and display them in map form for the first time, you need to give them credibility by displaying them with topography as the background layer in the form of a hillshade dataset. With the basin boundaries layered on top of the hillshade, the viewer will immediately see how the basin boundaries line up with the ridges on the hillshade. Additionally, regardless of whether or not your basins are the central feature of the map, you usually want to include a hydrography layer with the basins. Because basins are tightly coupled conceptually with streams and rivers, your audience will be keen to see them in relation to your basins.

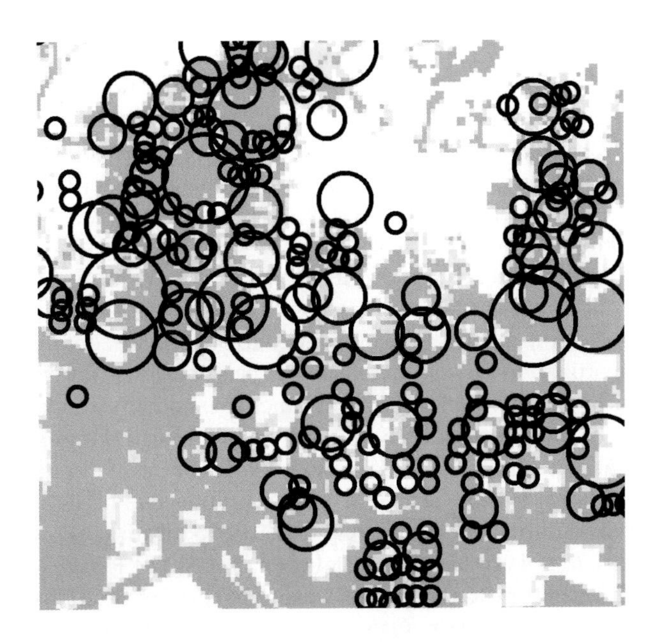

FIGURE 6.72
The same impervious surface data and color scheme as in the previous example form the background in this example. They have been lightened significantly to accommodate the addition of another dataset: number of households per census block.

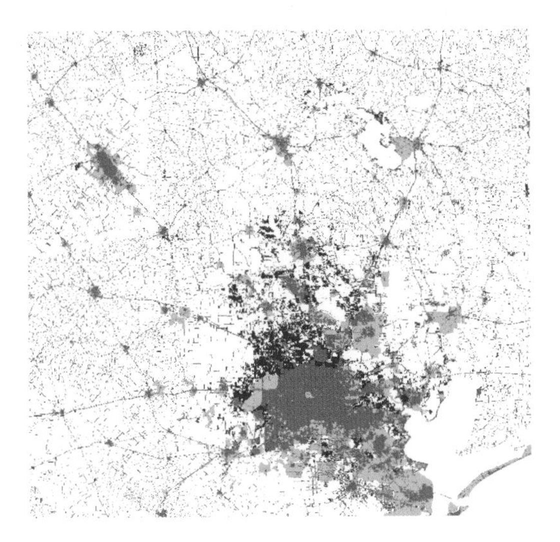

FIGURE 6.73
For this medium-scale view of the NLCD percent impervious data, all the impervious pixels are the same color—red—so as to simplify the visual correlation between the populated places (transparent gray) and the imperviousness. With just this simple map, a viewer can start to see that impervious surfaces not only cover urbanized areas but also connect urbanized areas to one another.

Bold red is a good color for basins since it likely won't compete with other linear feature colors and is striking enough to bring attention to them. Purple is another good choice since it is close to blue on the color wheel. Blue, of course, is the color for hydrography, and basins are closely related to hydrography. Any bright colors would be good as well. It is usually not necessary to use a fill color or pattern on a basin map as that would only obscure the features in the interior of the basin. However, if your map's focus is solely on the basins themselves, without any detailed background data, you could borrow the fill technique discussed in the "Political Boundaries" section. For this technique, you use a series of related colors as fills. The colors need to be different enough, especially for adjacent features, to be able to tell where each basin is. When your basins are nested within each other, a common practice is to use a thicker outline for the major basins and subsequently thinner lines for each tier of smaller basin size. Only use this technique if it is really called for because it could potentially create useless visual clutter. The color of the outlines can be lightened for the nested basins and darkened for the major basins. To minimize confusion, however, make sure to not change the actual color. Please see Figures 6.74 through 6.77 for examples of how to depict this feature in certain situations.

Occasionally, you will run into an issue with the basin scale and the other map element scales not being entirely coincident. When the basins are of a finer scale than the rest of the map data, they can wind up looking too chunky due to having many more nodes than needed at a small scale. To fix this problem, you will have to get rid of the excess nodes by

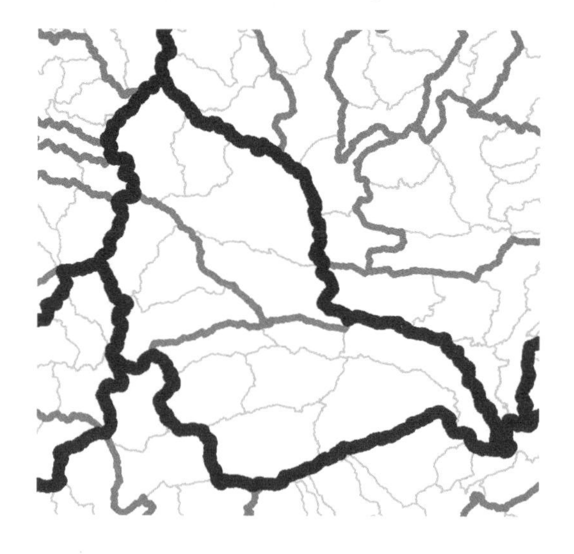

FIGURE 6.74
These nested basin levels are mapped with varying shades of red and varying line thicknesses. The smaller-sized basins (subtributaries) are the lightest shade of red and the thinnest line. The medium-sized basins (tributaries) are a medium shade of red and a medium thickness line. The large-sized basin (mainstem) is the darkest shade of red and the thickest line. (This could be referred to as the porridge principle, à la Goldilocks.)

FIGURE 6.75
The basin in this map appears to be floating. The absence of data surrounding the irregular basin outline focuses the viewer's attention on the basin. This is a great technique to use when you are lining up a series of maps for comparison such as a time series. It is not effective for maps where the viewer needs to have an understanding of the surrounding terrain.

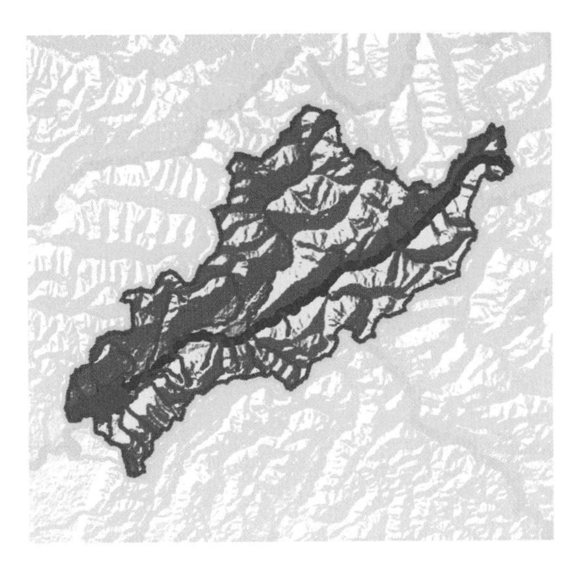

FIGURE 6.76
If you want to achieve a balance between providing the viewer with contextual detail and focusing the viewer's attention on one particular feature, use this semimasked technique. To produce this map, I simply selected the basin and hydrology that I wanted to highlight, made them into new layers, and created lighter symbols for the focus area and darker symbols for the outside area. Some GIS programs allow symbol levels, which can be used for this effect as well. You can also create a transparent layer that covers the outside area to mute any of the data under it.

FIGURE 6.77
A hillshade map is a natural accompaniment to the basin delineation shown here. The hillshade is much too dark in its default form to allow other data to be seen on top of it. This hillshade was lightened by about 50% and the basin color was changed to yellow, which acts as a highlighter on the dark background.

generalizing the data, finding coarser-scale basin data, or creating your own coarser-scale basin delineation with a course elevation dataset. On the flip side, when the basins are of a coarser scale than the rest of the map data, they can look jagged due to having many fewer nodes than the surrounding data or, even more egregious, have streams that actually cross them (which is antithetical to the definition of a basin). In this case, your options are to accept the scale inequalities, call out the discrepancy as a caveat in the text portion of the layout, find a finer-scaled basin delineation, or create your own finer-scale basin delineation with a high-resolution elevation dataset.

Buildings

Colors*

Building footprints are planimetric polygons or polylines that delineate the outer edge of the first floor of buildings. Usually they are digitized from

* The color bar contains suggested colors for buildings.

aerial photographs, either by hand or with feature extraction software, but they can also be delineated via Global Positioning System (GPS) or architectural plans. Sometimes the data only contains the geometry, but it could also include feature attributes such as building type (residential versus commercial, for example), number of floors, construction material, square footage, and so on. Building footprints, while used for a variety of purposes, are especially well suited for emergency management tasks such as color coding buildings according to which fire hydrant is nearest and locating homes on large parcels.

Maps of building footprints are particularly well coupled with parcel lines, orthophotographs, and LULC data. Parcel lines show where within a parcel the building is located, where vacant parcels are, and typically provide a modicum of spatial context given that road, lake, and other landmark outlines are easily discerned from parcel line data. Orthophotographs provide even more detail concerning landscape features, and consequently the spatial context for the footprints, but do not usually provide any information concerning the land property size and location, as parcels do. LULC data is helpful when you want the viewer to self-assess any relationships between landscape feature categories and footprint locations or sizes.

When parcel boundaries are used as the background layer, the footprints need to be shown as solid features in order to differentiate them from the parcel lines. However, when orthophotographs or LULC data is used as the background layer, the footprints are best symbolized as outlines so the complex background data can be seen through the footprints. Another symbolizing technique with this kind of data is to generalize it by eliminating all the features that aren't necessary. This allows a smaller-scale view to be used without the footprint features overwhelming the map due to their large quantity. For example, the residential building footprints could all be eliminated from the view in favor of displaying only the commercial building footprints. Another possibility is to show only building footprints that are within some distance of another feature, such as all the buildings serviced by a particular fire department or all the buildings within a stream buffer.

Recently, great strides have been made in the rendering of buildings in 3D software for better visualizations, either using stereo photos for data or modeled using procedural techniques. With stereo photos, the 3D information is built into the data and the buildings appear as they do in real life. With procedural techniques, 3D modeling software is used to create maps from 2D GIS data by determining building height from the GIS attributes and depicting the buildings in a realistic fashion by means of texture facades. This latter type is useful for creating fictional cities and realistic planning scenarios (see Figure 6.78). Both types of visualizations, which are generally shared as interactive maps, can be used for urban planning, gaming, and defense simulation, among other uses. Please see Figures 6.79 through 6.82 for examples of how to depict this feature in certain situations.

FIGURE 6.78
A procedurally generated 3D model of buildings in a historic Arab city center. Elliot Hartley, Garsdale Design Limited.

FIGURE 6.79
These gray building footprints are surrounded by lighter gray parcels. The parcel layer gives these buildings a lot more context, thereby adding to the viewer's understanding greatly.

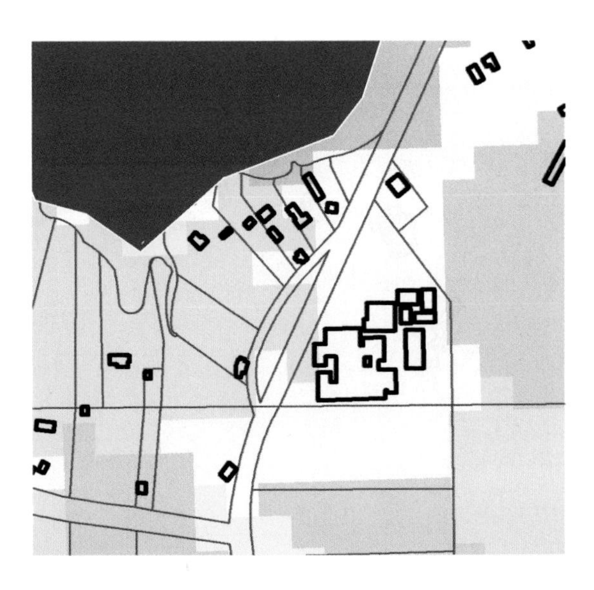

FIGURE 6.80
This map of land use and land cover also shows building footprints and parcels. The building footprints are outlined so that the land use land cover data can be seen underneath. A map like this shows that you have to be careful about scale when there are data like building footprints. The footprints could be off from the underlying data layer, if, for example, the footprint data is derived from 18-inch imagery but the land use land cover are derived from 30-meter imagery.

FIGURE 6.81
A high-resolution orthophotograph is the background for these building footprint data. The footprints are outlined so that the orthophotograph can be seen underneath them. The orthophotograph is slightly lightened and only those buildings within a 150-foot buffer of the stream are visible.

FIGURE 6.82
This simple building footprint map explores the relationship between the shoreline, parcel lot lines, and houses. The reddish color of the buildings allows them to become the focus of the map despite their smaller size in relation to the parcels and water. In this case, the bold color applied to small but well-dispersed features is very effective. If the buildings were larger, the bold color might overwhelm the map unless the other features were similarly bold.

Soils

Colors*

For this discussion, I use the terms *soil property* and *soil type* to mean two distinct things:

- Examples of soil properties: percent clay, rockiness, pH, drainage, electrical conductivity
- Examples of soil types: clay, loam, silt-loam, clay-loam

Spatial soil data describes soil properties at certain locations and depths. There are many potential soil properties and they are usually described in one or more fields of the soil data table(s) such as pH, texture, and water-holding capacity, among others. Soil data also includes one or more soil-type fields that utilize a standard taxonomic system that groups all of the properties into categories. There are several taxonomic systems in use around the world, usually standardized by country or region such as the US Department of Agriculture soil taxonomy and the World Reference Base for Soil Resources. Soils data is

* The color bar contains suggested colors for soils.

not standard 2D data. Because soil properties and types change depending on the depth of the soil, soil data will include property and type data, not just for the surface polygons, but for subsurface polygons as well. These different depth layers are known as horizons. This has implications for how the data is visually displayed on a 2D map. Usually vector based, soil data has a one-to-many relationship between the polygons and the linked horizon attributes.

The unique thing about soils is that soil properties are a continuously changing variable even though the data that captures them is usually a set of discrete polygons stretching across a geographic area. Adjacent polygons with different soil-type attributes may, in fact, be quite similar near their common edge despite their disparate categorizations. In other words, while soils are continuously changing in real life, the data that describes them is discrete in the database. Sometimes soil properties, and therefore soil types, change over a very small distance like a few centimeters and sometimes they change over a large distance like a few kilometers. This is because the mapping units, or polygons, almost always contain a variety of soil types within them. Only when you get to the very large-scale datasets, like 1:5,000 or larger, do the soil mapping polygons begin to identify single types of soil. For most national and regional datasets, therefore, the soil polygons will contain a one-to-many relationship to soil type. These soil types themselves change in concentration within their soil polygons.

The data is usually obtained via soil sampling at regular intervals, which is then interpolated to create the GIS soil data. The sampling and interpolation can occur over a very small tract of land, such as a farm field, to aid in precision farming analyses, or they can occur over a regional or nationwide scale in order to facilitate coarse-level correlation analyses or site-suitability analyses, for example. You should definitely become familiar with whatever standard soil taxonomy your data uses before attempting to make it into a map. This is important because you may have to manipulate the data via generalization or some other means in order to properly present the data in a readable manner.

Soil Datasets

There are two major soil datasets in the United States. They are both administered by the Natural Resources Conservation Service (NRCS). First, the US General Soil Map (formally State Soil Geographic Database, or STATSGO), is a coarse-level database of soils that covers the contiguous United States. This data comprises the first comprehensive soil mapping that the United States undertook, but has since been updated using finer-scale data that has been generalized. The finer-scale data is from the second major soil dataset in the United States, called the Soil Survey Geographic Data Base (SSURGO). This data is at a much higher scale—1:24,000—and covers most of the United States. There are other national soil datasets in other countries, and there are also innumerable soil surveys conducted at an individual parcel level (usually farms).

How to Show

Soils can be shown in a categorical color scheme or via choropleth mapping. Categories would, obviously, be keyed to the type of soil present in a particular polygon or cell, while a gradient color palette would be applied to, say, the percentage of clay as an average of the horizons or some other particular variable in the soils. Because soil maps could potentially have a huge amount of colors given the vast number of individual soil types, this is another good place to mention generalization techniques. But before I do, let's remember that doing so is dependent on the scale at which you are mapping. If you are mapping soils at the neighborhood, farm, or individual parcel scale, you will be able to show all of the colors needed and, indeed, you will most likely be required to, in order to provide your viewer with all the information possible. For these large-scale maps you may need to include spatial context by means of parcel data, building outlines, roads, or other defining features. If only general information is needed or if your scale is at a region or national scale, then it is likely that you will need to generalize this data in some manner.

How Many Categories Is Too Many?

A good guideline for the number of classes to show on one map is 10–12 with 10–12 separate colors. Though we are *capable* of distinguishing up to about 20 separate colors, in reality, our eyes become fatigued with that many colors. In order to avoid this fatigue and any errors that might stem from it in terms of your viewer's interpretation of the map, it is best to keep the number of colors to a smaller amount. Any more than 10–12 and you should employ some kind of patterning to the excess categories. Patterning, such as crosshatching, can easily overwhelm a map, however, so if you use it, be sure to apply it to those categories that comprise the least map area.

To keep the number of classes low, you will need to generalize the data in one or many of the following ways. First, you can simply use the built-in generalizing schema that is in the soil data attributes. Soil data attributes typically describe a hierarchy of nested soil types. Symbolizing the data on a higher order in the hierarchy can lower the number of classes significantly. Second, you need to account for the soil horizons. To do that, you can calculate the average of all depths with regard to an attribute such as permeability, or you can display only the maximum rockiness, minimum electrical conductivity, and so on. Third, you can dissolve soil mapping units to create larger units and perform the same kind of algebra to create aggregated soil units such as average or maximum.

As with many of the other feature types discussed in this chapter, soils data can be depicted via isolines instead of the traditional polygon or grid format. To create an isolines dataset, you need a point layer of the original soil survey locations as opposed to an interpolated layer. The GIS is used to connect these points according to a common property. For example, a map of soil pH could be created by connecting points of equal pH to create a line layer of soil pH. In some cases, this type of soil layer could be more helpful than a polygon layer of soil pH if, for example, you want to overlay the soil pH onto another a layer such as an orthophoto, elevation, or LULC.

Another way to display such complicated and fine-level data on a map is to create a dot map. These have the benefit of having less generalization and thereby allowing more of the data to be visualized. Dot maps are increasing in popularity among geoprofessionals and it definitely behooves us to have them in our arsenal of skills. A particular kind of dot map for soils was developed by Linda Barrett, who is an associate professor of geography and planning at the University of Akron. The great thing about her maps is that they allow the spatial continuum of soils to be more visible on a printed map than the more traditional, single-variable, choropleth map. Often times we display a soil map by showing only the dominant soil type within each soil mapping unit even though the data contains information on the percentage of each soil type within each soil mapping unit. To give people an understanding of those percentages and how they vary across the dataset, these dot maps show the percentage of each soil within the unit. This is accomplished by placing points randomly within each mapping unit, color coded by soil type, where each point represents a certain percentage of soil type for that unit. Please see Figures 6.83 through 6.85 for examples of how to depict this feature in certain situations.

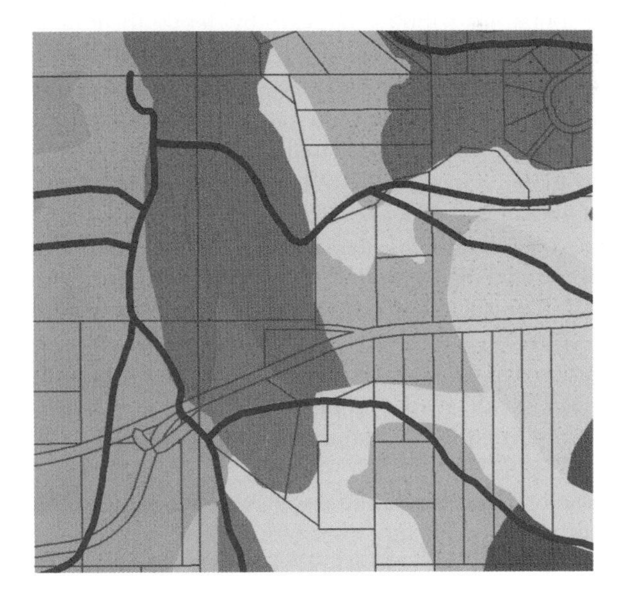

FIGURE 6.83
No, it isn't a groovy t-shirt from the 1970s. This is a soil map underneath hydrology data and parcel boundaries. Since soil is a continuous variable, this map deemphasizes the boundaries between class types by not using outline symbols.

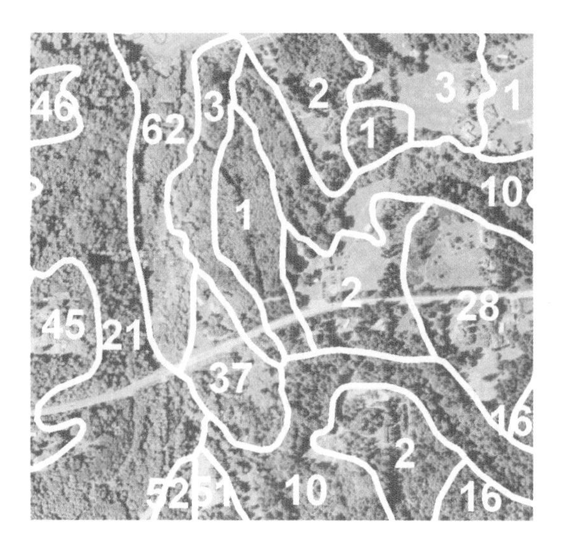

FIGURE 6.84
Because this map displays two overlapping polygon datasets, the soil data is shown as outlines and labels only. If you tried to use a transparent fill symbology for the soils polygons, the colors of the orthophoto would mix with the colors of the soils polygons and not create a cacophony of colors, none of which would match the map's legend colors.

From 2D to 3D (and in between)

Since soil data contains attributes that vary by depth, the data is better visualized in a 3D environment. However, most soil maps are still 2D due to limitations in hardware and software and traditional expectations. You can try to create a 2.5D map by draping a transparent soil map over a hillshade and, for example, visually exploring any relationships there might be between average permeability and elevation. Of course, the hillshade draping technique still does not allow any visualization of attribute differences between horizons and can only display averages or surface conditions. Because of that, the 2.5D map only gives a 3D *effect* as opposed to being truly three-dimensional. Exporting your data to 3D software or utilizing 3D GIS software is the only way to obtain a truly 3D map.

Combining your soil data with other data types, such as vegetation, can make an already complicated map much too complex for anything to be seen properly. If this kind of overlay is needed, then one of your best bets is to go ahead and combine the data yourself into a new dataset of combination categories and display those.

By now you've realized that soil data is some of the more complicated data to master. Before attempting any kind of mapping, let alone analysis, of this data, read more about it in general and certainly read more about your particular soil data in detail. Things can get complicated pretty quickly and you could inadvertently misrepresent the data. Remember that you hold the power to inform or misinform your map viewer; understanding your soils data is the key to making it the former and not the latter.

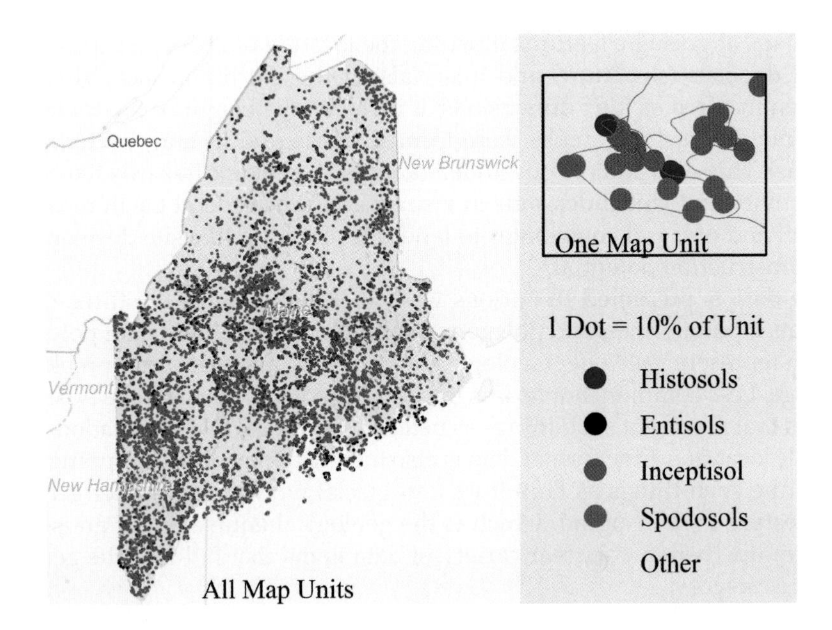

FIGURE 6.85
This soils map uses a dot-density technique. Each soil map unit is comprised of several types of soils (shown here are the four main soil orders in this particular taxonomy) that are shown as randomly placed dots in the units depending on the percentage of the unit that each comprises. So if a soil unit is made up of 70% spodosols and 30% entisols, then it will have seven spodosol dots (brown in this case) and three entisol dots (blue in this case). This data is from the NRCS STATSGO database map unit polygons and joined with the "comp" table from the same data source. The size of the mapping unit affects the visual outcome of a dot-density map as you can see in this example. Both large and small soil units contain 10 dots, so the larger units have more blank space. Of course, the dots in the smaller units are squeezed together. You can change the symbol sizes and the number of percentage points each dot represents until an optimal compromise is made.

Geology*

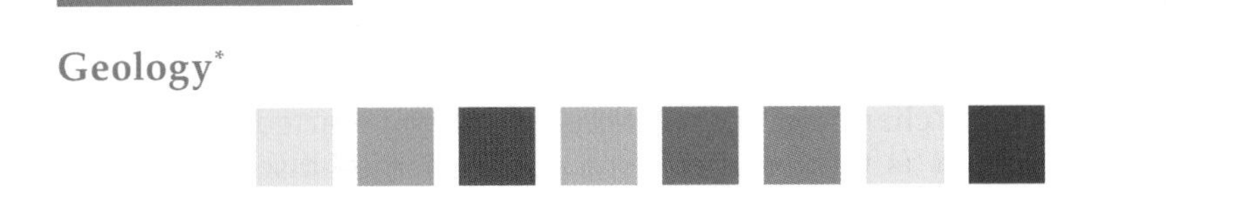

COLORS

It is anticipated that digital cartographers will be continually challenged to develop new techniques as software evolves and as geologists and users demand more complex and informative products.

US Geological Survey, National Cooperative Geologic Mapping Program[7]

* The color bar contains suggested colors for geology.

GIS maps of geologic features illustrate the location of rocks and unconsolidated deposits (structures and materials) at the Earth's surface. This data is often in three or four dimensions; it includes the height of the rocks and the time at which the rocks were formed (measured in eras, periods, and epochs). This data adds to our understanding of geologic hazards like volcanic activity and landslides, aids in visualizations of general Earth resources like oil and ore, and contributes to land use planning like site development and construction potential.

The data is presented in various ways including any of the three vector formats—points, lines, and polygons—with the most typical being polygons, which represent *map units*. Geologic map units are areas of similar rock type and age. Less common, but no less important geologic map feature types, are points that represent such things as bedding attitudes, fold orientations, and sample locations. Line format data is also included in geologic mapping, representing such things as fault lines and glacial moraines. Another geologic line feature is the isograd, which is the geological equivalent of an isoline. To sum up, there are a great variety of data items that fall into the geology feature category.

As mentioned above, rock height and age are often included in geologic spatial data. These enable geoprofessionals to create 3D geologic maps and indeed, they are increasingly being visualized in 3D because it is much more helpful to the end goals of the scientific audience for this feature type than a simple flat map. If you are, however, displaying the geologic information in traditional 2D, you may want to supplement this with one or more cross sections to illustrate subsurface feature trends. These cross sections, which are common on 2D geology maps, are always referenced in the main map via some sort of graphical construct, typically a dashed or solid line.

Symbols for such geologic mapping mainstays as landslide locations, faults, volcanic features, and land subsidence are somewhat standardized. The Federal Geographic Data Committee (FGDC) Digital Cartographic Standard for Geologic Map Symbolization contains a cartographic specification that you can use when you are putting together your geologic map.[8] The standard describes what symbols to use for certain features, and contains color charts and other items to get you started. The FGDC standard is known as the American standard. There is also an international color standard from the International Commission on Stratigraphy that is somewhat similar, though it differs for the Mesozoic (blue) and Paleozoic (green) eras.

However, even given those standards, choosing the right colors for geologic map units can be difficult, especially on maps that contain a large number of different map units. The general rule is that younger map units are assigned light colors while older map units are assigned dark colors. This kind of color scheme can help to illustrate the evolutionary construction of an area by allowing the viewer to see how the age of rocks changes through the landscape at a single glance. The Washington Department of Natural

FIGURE 6.86
The figure shows the colors used by the Washington Department of Natural Resources Division of Geology and Earth Resources for rendering map units by rock age.

Resources Division of Geology and Earth Resources uses the general guidelines in Figure 6.86 for rendering map units by age.

These colors mostly follow the colors found on the US Geological Society's Rocks of Ages legend, though the Quaternary period colors differ.[9] It is important to note that these colors apply mostly to sedimentary rocks. Igneous and metamorphic rocks—the other rock classes—are traditionally red and orange and brown and olive, respectively. It is also standard for geologic map unit legends to be organized with the youngest map unit at the top of the legend and the oldest map unit at the bottom of the legend.

Though there are definitely traditions and standards as discussed and illustrated above, these are not applied as consistently as those of us who want a simple key would hope. For example, some sources say that plutonic rocks are pink and volcanic rocks are red, while others say that plutonic rocks are red and volcanic rocks are orange. Even the concept of colors getting darker as the age of the rock increases is not equally applied everywhere. The best thing to do when handling new geologic data is to first find out if the originating agency has a predefined color scheme. If so, take a look at it and decide if it is adequate for your particular map. Pay particular attention to the differentiation of the colors between adjacent polygons. You may have to tweak colors here and there depending on the geology of the area you are mapping, so that differences between map units are easier to detect.

This issue also comes up when you are designing your own color scheme. While you might want to assign a different color for every different type of rock, of which there could be 50 or more types, you may not know until you have assigned all the colors whether or not two very similar colors will wind up adjacent to one another. When this occurs, you may need to consider an alternate color scheme or utilize patterning for one or more of the rock types.

Assigning a pattern fill to rock types is a common practice for maps when there are simply too many map units for color alone to allow adequate differentiation. However, as soon as fill patterns are introduced on a map, it will start to look cluttered. Therefore, they need to be reserved for situations where no other possibility exists for differentiation and they should be assigned to the map units with the smallest possible total area. Patterns will also tend to obscure any line features that are layered on top of the geology map units, such as contour lines, labels, and points, so keep that in mind when reviewing the finished map. A pattern guide for lithology is found in the FGDC Digital Cartographic Standard for Geologic Map Symbolization.[10]

A commonly cited peculiarity about geologic maps is that the audience for them is both the lay public and geological scientists. This means that though the maps must be highly detailed, they must also be understandable, at least to a certain extent, by a nonscientific audience. One of the things that is being done to improve the dissemination of geologic information to nongeologists is to standardize the maps even more than they already are, so that a map viewer does not need to relearn symbols, colors, and labels each time a new geology map is issued.

As far as completing your geologic map goes, you will be keen to know what other information is necessary to show on the map. Because things could get cluttered fairly quickly by adding new map layers, you will naturally want to limit the number of map layers as much as possible while still providing the necessary interpretive information. To that end, you may want to include hydrography, which, since it is a linear feature type, does not threaten to clutter the map very much. The main purpose of adding a hydrography layer is to aid in a viewer's understanding of geologic features since rivers and streams often follow geologic fracture lines. Please see Figures 6.87 through 6.90 for examples of how to depict this feature in certain situations.

Another layer to consider including in your map is elevation. Elevation is relevant to geology for many reasons. One is that it is often a factor in the delineation of ecoregions when combined with geology layers. Another is that it helps to identify areas that are prone to landslides. For example, some ecoregions are subdivided based on elevation in mountainous areas. Another is that a steep slope combined with permeable sands and gravels on top of silt and clay could indicate a landslide hazard area. Though it seems to be an important companion to geology, elevation data could easily overwhelm a geology map. If the elevation were itself depicted in varying shades of color just as the geology map, there would obviously be too many colors to decipher. One of the most common solutions for large-scale maps of this type is to use contour lines on top of the geology to denote the elevation. At medium and small scales, a transparent geology layer on top of a light hillshade can give the overall idea about how the two are related without trampling on the geology symbology unduly.

Yet another layer (or set of layers) to include with the underlying geology map units are the accompanying point and line geologic features. When placing these on the map, you must ensure that they stand out from

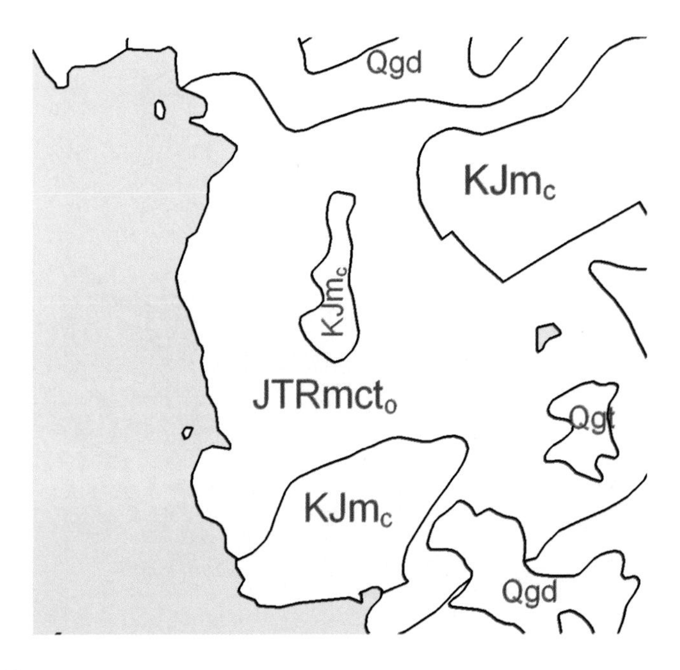

FIGURE 6.87
This very simple geology map shows the map units in white, labels in dark gray, and water in gray. If a black-and-white map is your goal, you might be tempted to use gray for the land and white for the water. However, in this case you can see that the white actually pops out more at the viewer than the gray and also gives a lot of contrast to the dark gray soil labels.

FIGURE 6.88
Using a standard color scheme, this Washington State map of geologic age shows where the older rocks (darker colors) are and where the younger rocks are (lighter colors).

FIGURE 6.89
A view of Mount Rainier, Washington, contains a muted geology layer in the background so that the overlying layers (national park boundaries, hydrology, and roads) are highly visible.

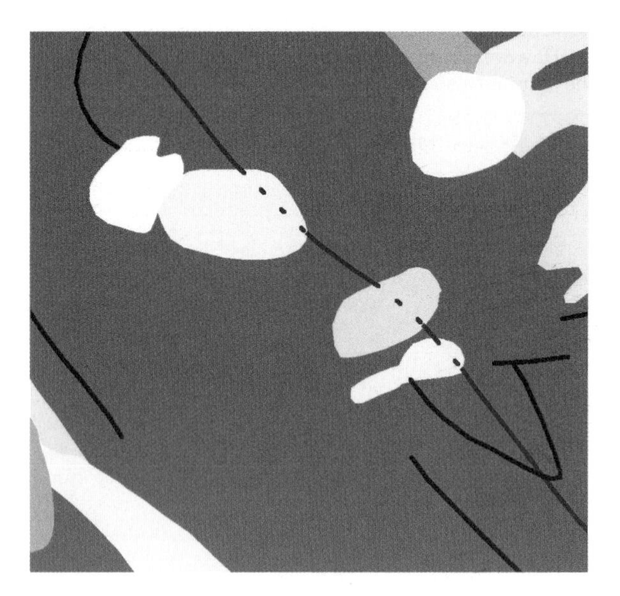

FIGURE 6.90
These geologic line features, shown in purple and red, are overlaid onto the geologic map units. The line features are bright since they can easily become lost within the colorful map unit polygons due to their small size. Line features such as the dikes shown here can be differentiated by their vertical location with dots to denote concealment and solid lines to denote surface dikes.

each other as well as from the underlying map unit colors and patterns. To that end, faults are traditionally in bright red, dikes in bright pink, and so forth.

If map labels are warranted on your geology map units, you can make the map look even more complex. Labels are typically based on the codes that are used to identify the age and lithology (type of rock) for each map unit. Sometimes an additional part of the code includes an abbreviation of the map unit name or a note. The typical code consists of two Greek alphabet letters to denote the age followed by two small letters to denote the lithology, and finally an optional subscript to denote the name or note. Since subscripts cannot be put into a database field, these will often be put in parentheses or a separate field in the database. You will have to change these to subscripts when it is time to label the map if they are deemed necessary for the map's purpose. An example geologic map unit code would be as follows:

KJm(c):

KJ = Cretaceous-Jurassic age
m = marine sedimentary rock-type
(c) = Constitution Formation, Decatur terrane

Endnotes

1. See *Nautical Chart Symbols*, Chart No. 1, Version 1.0 (Abbreviations and Terms, Department of Commerce National Oceanic and Atmospheric Administration and Department of Defense National Imagery and Mapping Agency, Office of Coast Survey) Washington, DC: November 1997, http://www.nauticalcharts.noaa.gov/mcd/chartno1.htm (accessed January 29, 2014).
2. Bump mapping is an advanced visualization technique related to hillshading, borrowed from the broader world of computer graphics. Essentially, it is a way of providing additional realism by shading individual pixels or groups of pixels to represent the elevation of a part of an object or whole objects instead of the typical GIS map bare-ground elevation. A good rundown of the technique using the example of shaded relief for particular modeled forest stands was developed and written by Jeffery S. Nighbert, Department of the Interior, Bureau of Land Management, *Characterizing Landscape for Visualization through "Bump Mapping" and Spatial Analyst*, Esri User Conference Proceedings, Vol. 3, 2003, http://gis.esri.com/library/userconf/proc03/p0137.pdf (accessed October 12, 2013).
3. R. F. Uren and A. Coates, "Mapping the Human Body," *Government Technology* (1997), http://www.govtech.com/magazines/gt/Mapping-the-Human-Body.html (accessed October 12, 2013); P. J. Kennelly, "Not Mapping Our World," *ArcUser* 10, no. 3 (2007): 68–69, http://www.esri.com/news/arcuser/0807/nongeo.html (accessed October 12, 2013).

4. J. R. Anderson, E. E. Hardy, J. T. Roach, and R. E. Witmer, *Land Use and Land Cover Classification System for Use with Remote Sensor Data*. Geological Survey Professional Paper 964 (1976). A revision of the land use classification system as presented in US Geological Survey Circular 671. Washington, DC: United States Government Printing Office, http://landcover.usgs.gov/pdf/anderson.pdf (accessed November 25, 2013).
5. For an example of this process, see T. Patterson and N. V. Kelso. "Hal Shelton Revisited: Designing and Producing Natural-Color Maps with Satellite Land Cover Data," *Cartographic Perspectives, Journal of the North American Cartographic Information Society* 47 (2004), http://www.shadedrelief.com/shelton/c.html.
6. A. Comber, P. Fisher, and R. Wadsworth. "What Is Land Cover?" *Environment and Planning B: Planning and Design* 32 (2005): 199–209.
7. US Geological Survey. *Cartography & Geologic Maps* (2005), FGDC Digital Cartographic Standard for Geologic Map Symbolization (2006), http://ngmdb.usgs.gov/fgdc_gds/geolsymstd/fgdc-geolsym-all.pdf (accessed January 29, 2014).
8. See Section 5 in the introductory text and Sections 33 and 27, *FGDC Digital Cartographic Standard for Geologic Map Symbolization*, Federal Geographic Data Committee (Doc. No. FGDC-STD-013-2006) US Geological Survey Techniques and Methods 11-A2, http://pubs.usgs.gov/tm/2006/11A02 (accessed December 10, 2008).
9. US Geological Survey. Rocks of Ages: An Explanation of the Legend. A Tapestry of Time and Terrain: The Union of Two Maps—Geology and Topography, http://tapestry.usgs.gov/ages/ages.html (accessed August 3, 2008).
10. From the Pattern Chart in *FGDC Digital Cartographic Standard for Geologic Map Symbolization* (*Postscript Implementation*), Federal Geologic Committee (Doc. No. FGDC-STD-013-2006), US Geologic Survey Techniques and Methods 11-A2, http://ngmdb.usgs.gov/fgdc_gds/geolsymstd/fgdc-geolsym-patternchart.pdf (accessed December 10, 2008).
11. Wallkill Color. Munsell Software Conversion Program, Version 2014, http://wallkillcolor.com.

Suggestions for Further Reading

American Standard Geologic Age Color Scheme developed by the US Geological Survey and the Association of American State Geologists is shown in a reference diagram on this site. http://www.peoplelandandwater.gov/usgs/usgs_11-26-07_usgs-association-of.cfm.

A Tapestry of Time and Terrain shows the American Standard geologic rock age color scheme with hillshading underneath. http://www.tapestry.usgs.gov/ages/ages.html.

The National Geologic Map Database is maintained by the US Geological Survey and contains information on mapping techniques and guidelines. http://www.ngmdb.usgs.gov/Info/home.html.

The International Commission on Stratigraphy posts their stratigraphic color charts online. http://www.stratigraphy.org.

The Soil Geographic Data Standard, developed by the Federal Geographic Data Committee (FGDC), describes the standards for mapping soils in the United States with specific regard to the soil data created by the National Cooperative Soil Survey. http://www.fgdc.gov/standards/projects/FGDC-standards-projects/soils.

The Utilities Data Content Standard, created by the Federal Geographic Data Committee, is a good resource for looking up the standard attributes of many utility features. http://www.fgdc.gov/standards/projects/FGDC-standards-projects/utilities.

Study Questions

1. What is spot height labeling?
2. What are road shields? Find vector files for two common road shields and include them with your answer along with a note providing the road type and country or state for which they are applicable.
3. There are many examples of generalization in this chapter. Explain two of them.
4. When might you use bathymetric data on a map? Give two examples.
5. What problem occurs with political boundary thematic maps in terms of differentiating polygon outlines by color? What is the solution?
6. In the "Fuzzy Features" section there are several examples of datasets with imprecise edges. Name an example of a dataset with fuzzy boundaries that *isn't* listed and describe how you would symbolize it to indicate that it doesn't have a precise edge.
7. Briefly describe a hillshade. Find an example of a hillshade map online, include a small screen shot in your answer, and cite where you found it.
8. What is an isoline? Describe two isoline types found in this chapter. How are they typically symbolized?
9. What colors are used in temperature mapping and what do they denote?
10. What is the challenge with trail data symbolization and why is this an issue? What is the solution?
11. Impervious surface data is typically organized in one of two ways in the database. Describe both ways in a few sentences or less.
12. Briefly describe two unique aspects of soil data that can make its symbolization complicated.

13. How many colors is the human eye capable of telling apart on the same map? How many colors are more reasonable for a map?
14. Name a couple of ways you could lessen the number of colors used in a class-type map.
15. What is the difference between the terms *land use* and *land cover*?

Exercises

1. Make a map (can be static or multiscale, paper or digital) that showcases one feature type from this chapter. Find a dataset, choose a scale or set of scales (small, medium, or large) that is appropriate for the data, symbolize, and label according to principle described here. If there is another dataset that should be shown with the data, then do so but make the layout as simple as possible: include a title, map, legend, authorship text, and data citation(s). Float those elements over the map. Places to look for data include Natural Earth, OpenStreetMap, or a local government website.
2. Find a hydrography dataset at the country level. Symbolize it with three levels of hierarchy. Make a map of it, with only country borderlines for context. Include a title (e.g., "U.S. Rivers") and a citation. Make sure the level of generalization is appropriate for the scale. For example, at the US scale, don't include minor rivers, just the major rivers and their tributaries. The hydrography dataset can be for whatever country you wish or the country for which you find data.
3. Select a good font for place labels. Create a style hierarchy for country name, state name, city name, and town name. Write the four labels in the font you selected and in the style you came up with using a plausible example (e.g., United States, Colorado, Denver, Steamboat). Use any of these tools: character spacing, font size, all caps, different colors, bold, italics, color saturation for the necessary hierarchy.
4. Utilities data is usually highly codified in terms of symbology. Look up two types of utilities data (power lines, manholes, sewers, etc.) and a standard symbol for each. Include screen shots of the standard symbols that you find.

7

Static Maps

Hey, believe it or not, some of us still print out maps from time to time! What are some of the things to keep in mind when designing for static map media? How will your map be viewed, on a digital device or on paper? What image format would be best for exporting? Will it be printed on a large-format plotter? Will it be saved as a JPEG for a slide presentation? What resolution and other exporting options ought to be considered? The answers to these questions about static maps influence map design from the very beginning. That's why this section of this chapter, which deals with these formatting and exporting types of questions, is good to read through at the beginning of the map design process just to get an idea of what things to look out for in general. With the general concepts safely stowed in your memory, you can then refer back to this chapter for specific design and export tips while you are knee-deep in a map product design.

This chapter focuses on static maps. That is, maps that can't be clicked, zoomed, or panned. A static map can be a paper map—of course—but it can also be a digital map as long as it's just a static image. Static maps require knowledge of export resolution, export formats, and workflow with regard to geographic information system (GIS) software and graphics software. It is also important to think about how the static maps are going to be presented during the design phase, and to that end, a discussion of a few of the most common uses of static maps are outlined here along with relevant design considerations. These common uses are slide presentations, poster presentations, and reports. The aim is not to be completely comprehensive with regard to slide, poster, and report style, but rather to touch on cartographic consequences for each of these, and to pique your interest for further learning and experimentation in these areas.

DPI

DPI, which stands for the dots-per-inch ratio, is often confused with resolution. The resolution is what we are referring to when we say that a map is, for example, 100 pixels by 120 pixels in size. You might see the resolution written as 100 × 120. In this example, the map contains 100 pixels along the horizontal axis and 120 pixels along the vertical axis. However, this alone does not tell us how big the map is. To get at that, we need to know the size of the pixels. Since pixels are measured in dots per inch, let's say our 100 × 120 map has a DPI of 100. This would make the map 1 inch by 1.2 inches in size. The higher the DPI, the more dots within each pixel and the finer the resulting image, if the resolution stays constant. So a map with DPI of 200 for the 100 × 120 map would be 0.5 inches by 0.6 inches in size. Now, the reason we geoprofessionals often think only in terms of DPI is because our static map's size is a variable that we set at the beginning of a project and do not change come print time. Therefore, when it comes to changing a parameter to try for a smaller file size or, conversely, a finer-looking image, we either increase or decrease the DPI.

With all static media, you have to consider the output DPI at export time if your software allows you to specify it. If file size and export file-processing time are not issues for you, then you may only be constrained by your printer's output resolution. If you are printing on a simple color office inkjet that only prints at 300 to 600 DPI, then obviously there is no need for you to export to anything higher than that. For example, a printer with a maximum 600 DPI printing capability will print 600 little dots of image information in each inch and no more. So, check your printer's DPI output prior to specifying an output DPI if you are printing the map. Generally, printed materials need to be exported at 300 DPI at a minimum to obtain a quality print.

Where JPEGs, PNGs, and certain other compression-based file types are concerned, there is an extra variable to consider. These file types are made with a compression algorithm that lowers the file size by resampling the image to a lower quality. If you want a higher quality than the typical JPEG compression allows, but still want a JPEG file type for whatever reason (as opposed to simply switching to an uncompressed file format), then your software may provide you the option (sometimes only programmatically) to change the compression by changing the resampling ratio. Therefore, it is possible in some cases to get a JPEG that is not compressed or less compressed than it typically would be. Of course, you would have to contend with the higher file size and longer export processing time just as with the other uncompressed file types if you go this route. If you must have a JPEG, though, and need smooth, nonjagged lines for a professional appearance, then this may be your answer. In particular, maps with high contrast or a lot of sharp edges (i.e., the typical GIS map) do require a high quality in the larger map sizes. Usually JPEGs and PNGs are used for slide presentations, websites, and other media to be displayed on digital devices, which don't require high resolutions but do require small file sizes. These are ideal for that purpose.

Higher-resolution files require more export time and produce larger file sizes. We're not talking about a linear increase in time and file size, either. We're talking an exponential increase since the resolution is a two-dimensional situation as opposed to one. Doubling the DPI produces an output that takes four times as long to export and is four times the file size.

Export Formats and Workflow

GIS software and graphics software offer many file-type options for exporting static maps. Uncompressed TIFF (Tagged Image File Format) files and EPS (Encapsulated PostScript) files provide higher image quality than JPEGs and PNGs (portable network graphics) in most cases, but have large file sizes. JPEGs and PNGs are compressed, allowing for smaller file sizes, but with the compression comes a trade-off in quality. The compression process creates distortion around the edges of large features, such that they turn out looking fuzzy, with some little dots of the color drifting off into the surrounding colors. Sometimes this isn't discernible to the eye, especially when you have many small features (especially in the case of photographs), but other times it is quite noticeable. A good solution for quality and file size issues is the Adobe PDF (portable document format). These have small file sizes and good image quality, but are difficult to impossible to embed in other applications like slides or reports. Newly popular is the SVG (scalable vector graphic) format, a versatile, high-resolution format that can be manipulated in graphics programs easily and supports animation.

There are many cartographers and geoprofessionals alike who do not simply export straight from their GIS package to a finished map file for printing. This is especially true for those professionals who feel comfortable using Adobe Illustrator, Photoshop, Inkscape, or other graphic design and image manipulation software. The export files that are ideal for this process are PDFs and SVGs, though others can be used as well. The general workflow includes exporting layers or groups of layers from the GIS into the graphic design file format, performing aesthetic manipulations there, and then exporting to a finished image file for printing. It is also not uncommon to go back and forth between the programs to change settings, make different selections, and so on, and can therefore become quite an ordeal.

The advantages of this method of map design are that you have some more control over color, you can do fancy things like change individual letters in a text block to different styles, create outlined lettering and drop caps, generalize boundaries easily, render halos around features quickly (called "outer glow" in Adobe Illustrator and "glowing edges" in Photoshop), and so on. Sometimes it's simply easier to do things in graphics programs than the GIS. Old map files that are disconnected from their original projects or their original data can be changed in graphics programs if you have no other alternative and if they need just superficial editing like changing of colors or text. You can also easily change the entire look of a map using the color filters available in these programs.

However, despite the general usefulness of this workflow, it involves taking a map from a geographically centered product and putting it into graphic illustration software that knows nothing about how to line layers up in geographic space. This means that it is crucial for you ensure that your geography is preserved. One of the best ways to accomplish this is to place a bounding box around your map layers prior to exporting them from the GIS. Then you can use that bounding box to line up your layers in the graphic illustration software. You also lose access to the spatial database, so you are obviously confined to only fine-tuning the visual output as opposed to doing any analysis, dynamic labeling, attribute-based symbolization, and so on. Make sure that all attribute-based manipulations are done prior to exporting to the graphics software. Additionally, you may want to investigate a bridge software product that provides the capabilities of both GIS and graphics design such as MAPublisher.

Some specific tips for the GIS-to-graphics editor workflow include the following:

- Make sure raster files are on the bottom of the layer list in the GIS before exporting.
- Don't set any layer transparencies prior to export.
- Make sure the colors for each data layer are unique so you can select them easily in the graphic illustration software.
- Arrange the layers in the GIS in the order in which you want them layered in the graphic software.
- Export in the projection that you need for the final map.

Obviously you will need to learn a lot about the other specific quirks concerning export to graphic illustration software as well as learn the software itself in order to make this an effective method of map making. Obtaining at least basic-level graphics editor skills is important in our profession. Polished, professional products are a result of not just knowing what techniques are possible, but also knowing how to achieve those outcomes most efficiently.

Slides

Most of us will be called upon to give a slide presentation at some point in our careers. If you are an analyst who toils away ceaselessly at the computer, never seeing a presentation, let alone giving one, you might not be in need of these skills. But you can be assured that you will at least be called upon at some point to add some slides to a presentation deck or to create an entire presentation that someone else gives. Whatever your circumstances, these slide techniques will really put some cutting-edge skills into your repertoire. You will not only wow the usually stupefied presentation audience, but will also communicate your points effectively. And those are two very good things to try to achieve!

Presentation Style: Fast or Slow, Busy or Simple?

Depending on your audience and subject material, you could design a slide presentation that is fast and simple, slow and simple, or slow and busy. The only combination left out is fast and busy (many slides, a lot of information on each slide). In fact, most typical conference presentations are designed in the fast and busy format, and this is probably the worst possible combination. These fast and busy presentations are both boring and hard to follow at the same time. With too many words on each slide, the presenter loses the audience as the group tries to read through all the words, forgets what the presenter is saying, and still doesn't understand what the main point is!

Fast and Simple

I like to design slides in the Lawrence Lessig style, which is fast paced with minimal information on each page.[1] It is kind of like the report style introduced in the next section whereby you don't want to interrupt the text of the report too much with your map, so you just keep it simple and let it flow in line with the body text. The same approach can be taken with slides by incorporating your slides seamlessly into the audio portion of your talk. The audience needs only to glance at each slide in order to get whatever visual cue you are aiming for. I encourage you to take a look at, and study, some of these fast and simple types of presentations online. The good news is that more and more people are adopting this style since the first edition of this book was published, so there's now a good chance you've seen several good talks to emulate already. Lightning talks, now available at most industry conferences, should be in this format. You will find that these presentations take on a movielike quality that moves the audience through the presentation as if it were a story. The result is a rapt audience that remembers the points you set out to make.

Characteristics of a fast and simple slide show:

- One or two words maximum per slide
- No background graphics
- Minimal color on text slides (try black text on white background or bold white text on a black or dark gray background)
- About four to six slides per minute
- A lot of simple graphics
- Slides are highly choreographed with the verbal presentation

Slow and Simple

The slow and simple method uses some Lessig-style slides, but a lot fewer of them. This sometimes suits a GIS presentation when you absolutely have to display a map that will take more than 10 seconds for the audience to absorb. For this style of presentation, you can sprinkle one or two word slides (10 words maximum) in between the complex maps for visual relief and variety. You will have many fewer slides with this method than with the fast and simple method. For a 15-minute presentation, for example, you might have just 15 slides as opposed to 100 or more.

With the slow and simple method (and also sometimes with the fast and simple method), do not shy away from having a script written out to help you through the longer intervals when the map slides are shown. Believe it or not, many professional presenters will practice a presentation, word for word, from a prewritten script 10 to 20 times before the real deal. Practicing means reading it on your own and in front of (very patient) loved ones or colleagues. If you have given your script a trial run many times and in front of others, you don't need to worry about looking as if you are reading throughout the presentation because most of it will be memorized by then. Even if a few sentences wind up getting ad libbed during the actual presentation, the script will help out tremendously.

To write out your script for practicing, try printing it out in a handwriting style font, or if you have loads of time, write it out by hand. It is much easier to read through a handwritten script or handwriting style of font when in front of an audience. Be sure to print it out in a large font—14 point at least—and double space it so that it is easy to follow. You can cue yourself to slide changes by simply including the word *slide* in a different color than the rest of your text at each slide-change location.

Tips for Creating Simple Map Slides

Another application of the simple slides method—whether or not you are going to present fast or slow—is to create several map slides, each adding to the last, achieving a dynamic map effect. This avoids audience fatigue by enabling them to focus on just one added data layer at a time. It makes your slides simple as opposed to the typical busy multilayered GIS map. Adding the map layers as you advance through the slide deck achieves a dynamic quality because the layers "magically" appear as you refer to them (in reality you are pressing the next slide button). Don't confuse this with slide animation effects where the slides fly in at you or fade out; those can be very distracting.

Consider this example: when I was in front of an audience describing how a project had utilized impervious surface data in an analysis (Figure 7.1), and then wanted to point out that the reason the map didn't show any impervious surface in one large section was because it was a national park (Figure 7.2), I used two slides instead of one.

Now, when I advance from the first slide to the second, it doesn't look to the audience that I have changed slides, it simply looks as if I have added the national park layer onto the first slide map. If I show just one slide with both layers, I would have probably needed a pointer to get the audience to look

FIGURE 7.1
This was the first slide in a presentation that showed a project study area and its key data—impervious surfaces (in yellow).

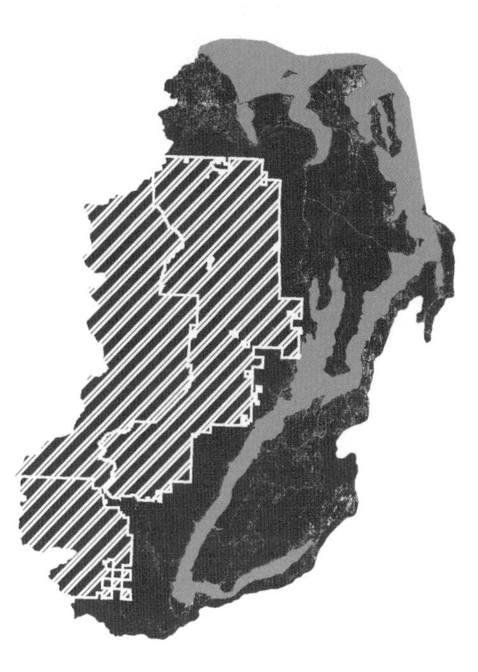

FIGURE 7.2
This was the second slide in a presentation that illustrated where the national park and national forest lands are within the study area.

at the little yellow dots when I said "impervious surface" since the national park and national forest lands layer dominated the rest of the map with its visual weight.

Note also that there is no text or legend on these slides because I simply told the audience what the first map was showing (impervious surface pixels in the Hood Canal Summer Chum watershed), and then quickly advanced to the second map with another sentence, "and here are the national park and national forest boundaries, so you can see why there was no impervious surface showing in that portion of the watershed." I wanted them to listen to me, not read words. The point of a visual is to augment the presenter's words, not to take the place of them.

Slow and Busy

While simpler is usually better when it comes to slide design (even though many presenters try to pack as much information onto slides as possible), there is one situation in which it can be alright to design complex slides. When you are presenting technical information to a small group of people who are already familiar with your project and who are there expressly for the purpose of learning the minutiae of your analysis, complex maps, tables, and graphs can be necessary. Do not think that a lot of text would be called for in this situation, however! Large text blocks and lengthy bullets are almost never warranted.

To give you an example of a complex slide, consider the following situation: The slide in Figure 7.3 was designed for a small-group presentation of fewer than 10 people. The presentation was held in a small conference room where the attendees sat around a conference table so they could react to and discuss the points as I spoke.

Since the presentation was more of a starting point for discussion rather than strictly a one-way information exchange, it seemed more natural to allow some complexity in the slide in order to convey the complexity of the analytical process we were discussing. Each slide was left on the screen for several minutes so that the group could study and refer to it as we discussed the issues.

Time and Research Wanted!

Remember that a good map slide can be made great if you take the time to consider your audience—not just their level of prior understanding of your material, but also things like audience comfort (affecting placement of maps on the slide), audience size, and desired outcome. Mind the color saturation—if possible, check to see how the map slide looks in the conference room right before the talk. Depending on the light level and the projector, the map may appear washed out. If the map is saved as an image file, you can do a quick fix by altering all the color saturations at once in your slide or image software.

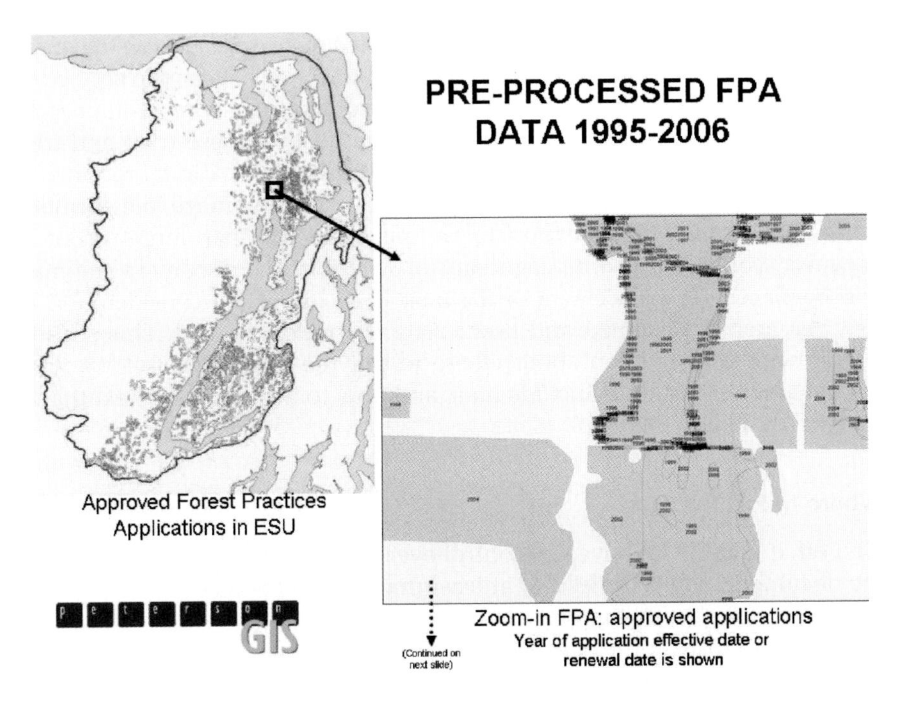

FIGURE 7.3
While this slide is too complex to show at a large-audience presentation, it is perfectly acceptable to show it in a small-group discussion concerning the intricacies of the illustrated analysis.

Also, be sure to mix it up a little. You need not adopt one method of presenting and stick with it for a decade or more. If you haven't done a presentation in a while and have an important one coming up, do some research on the latest presentation styles and slide designs. For example, at the time of this book's publication, YouTube videos online were a great source of ideas. You can watch other people give presentations that are similar in scope and subject material as yours and get ideas about the current trends, what to avoid, and what to include. As fashions change, there will always be new strategies for you to research and adapt to your individual circumstances. All of these tasks may seem time-consuming. They are! But your audience will thank you for it. Or at least remember it, which is way more than can be said for many talks.

Reports

Traditional paper reports were still a common way to disseminate the results of a GIS analysis at the time that the first edition of this book was published. However, in the intervening time, the report's popularity has diminished and it is now much less likely that you'll have to create one or create the maps for one. It's much more likely that you'll need to create a web map service or a static, digital, presentation instead. For those occasions when you need to provide an old-fashioned report though, take heed of these tried-and-true design rules, which apply to both paper reports and static digital reports.

Reports generally incorporate their static maps in image file formats. You might be tempted to export to an image file and then forget about it. However, you must also be cognizant of where in the document the map will be placed. The placement of the map is an important consideration in how the map is designed and how easily the report is read. This section begins with a discussion about where to put the map, then follows with what margin elements to include on it, and how to deal with any text that is specifically tied to the map.

Where to Put the Map

First off, if you do not have full control over the placement of your map(s) in the document, you will need to at least make suggestions concerning your map's placement in the report using the guidelines presented here. If you do have full control over the placement of your map(s), then you have the opportunity to integrate them in a professional and aesthetic manner. Some of the placement guidelines for small, in-line report maps include the following:

- Aim for a map size of about one-fifth to one-sixth of the overall page area, maximum, in order to integrate it into the document text (see Figure 7.4).
- Allow the text to flow around the map.
- Place the map near the text that references it.
- Do not place a map on a different page than the text that references it.
- Do not place a border (frame) around the map.

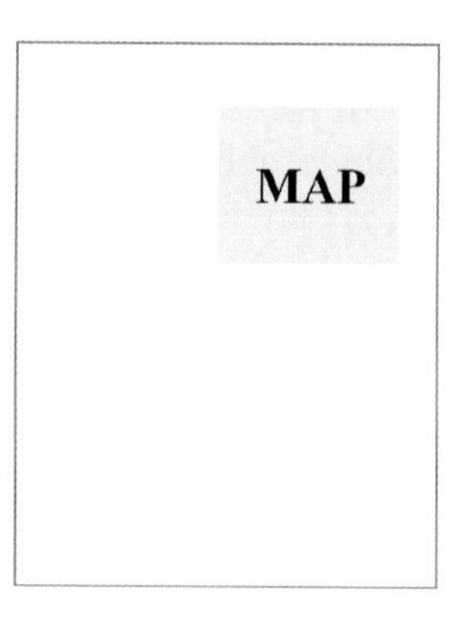

FIGURE 7.4
A good size for an in-line report map is one-fifth to one-sixth of the overall page area.

A map that takes up half the page is not the worst thing in the world, but if you can make it smaller, then do so, as it won't be so obtrusive. Also, if the map is not wide enough to stretch from one margin to the other, then do not float it between the margins surrounded by a bunch of extra white space. Simply let the text flow around it, magazine style. Often, when I'm showcasing cartographic examples on my blog, I'll screenshot them as rectangles that stretch from one side of the text area to the other, while being quite squat. The rectangular screenshots take on the same shape as a single text paragraph and thereby flow within the blog text perfectly. Regardless of size, don't put a frame or border around the map unless absolutely necessary. Allow the white space to be the frame so that the map and the text are the focus. Instead of neatening up the document, the border will tend to chop it up and cause distraction. This goes against the intuition of the non-design professional. If you are not convinced, simply try producing your report both ways: one with the map borders and one without. In most cases, the one without borders will look more professional.

Margin Elements for Report Maps

What margin elements could and should be included on a report map? Reports, especially of the scientific type, typically contain captions that are referenced somewhere in the body text and serve to explain the map or graphic without the reader needing to go back to the text for an explanation. Because of these captions, it is usually not necessary to include titles on the map, since that information is found in the caption. If you have access to the person responsible for the final report product, you will want to discuss whether or not captions will be included and even suggest captions when needed and if possible. Even without captions, a title is optional depending on the context. For example, I recently came across a US city zoning document that explained the various restrictions and permitted land uses for the zones in the city. The beginning of each new zone description was accompanied by a small map set off to the right-hand side of the page. There was one map per zone that showed where the zone was in relation to the rest of the zones. The maps had no borders, the text was wrapped around the maps, and they had no titles or captions, yet they were highly effective as quick reference graphics to give immediate context to the zone descriptions.

While a title may not be needed, the smaller in-line report maps occasionally contain scale bars and authoring information, and if it the features aren't obvious, a legend. The point is to keep the margin elements minimal so as not to disrupt the flow of the text and to keep the map simple enough that it can be deciphered at a glance. To accomplish this while still communicating some of the details that would have been put into the margin of a standalone map, put those details into the text of the report so that the reader is essentially "reading" the map as part of the text. Remember those stories from childhood that substituted pictures for words to aid early readers? The picture quickly illustrated the concept, you said the word, and then went on with the story. The same can be done with carefully thought-out maps in reports. The data must be generalized and the map element clutter-free in order to achieve such an effect, however.

When it comes to making a full-page report map, there is more design flexibility. This is due to the extra space and the stand-alone nature of the map. A title may be warranted if there will be no caption (e.g., if it is part of an appendix of maps). Legends are more important on these larger maps because the size enables more layers to be shown on them, and more layers means more deciphering on the part of the map viewer. Furthermore, especially in the case of an appendix map, a small amount of explanatory text placed on top of the map element can aid the reader in understanding how the map fits in with the report's content. For example, an appendix with 10 maps of an island's nearshore environment might have title text at the top of each page saying something like, "Nearshore Degradation: Map 1 of 10." This way, when someone prints the report and separates this map from the others, it will at least have some context written directly on it.

Do steer yourself away from creating elaborate map layouts on letter-sized report maps, though. There simply is not enough space and no need for the disclaimers, data source descriptions, and other supporting information that would go in the margins, since those items can more effectively be incorporated into the text of the report or in an appendix to the report. If the map is in an appendix, you can write an introductory page that goes at the beginning of the appendix and explains the maps and makes note of the elements listed above. If absolutely necessary, you can include a separate page as a legend for all the maps in a group if the legend won't fit on the same page as the maps. However, this is to be avoided if possible since you don't want the reader to constantly have to flip the pages back and forth in order to interpret the maps.[2]

When it comes to reports that are intended to explain the methods and results of a complicated analysis, it can be a struggle to fit everything on one page in a readable manner while still managing to display all of the relevant data. It can be very difficult to convey the magnitude of an analysis to readers when all you have is one (or even a series of) 8.5-inch by 11-inch pieces of paper! For example, I will often conduct analyses on a subbasin scale across a broad geographic area. In some cases there will be a couple of hundred subbasins with many attributes to display. It is difficult, yet important, to show all 200 subbasins on one sheet, so the reader can view the results at a glance. Usually the compromise is to use a color gradient (kept at five colors or less) or grayscale gradient to display one variable across the geographic area. Sometimes this means that a particularly small subbasin that contains the highest value or lowest value is not visible. In those instances, the small subbasin's value can be highlighted with an inset box that enlarges that particular area or a label with the value connected to the basin with a leader line. The map is generally accompanied by a table of the basins and their attributes in the report. I find it is important to include that information, even if it is also being supplied in digital spreadsheet and GIS form, because many of my potential readers will not have the time to view the digital data and will rely, instead, on the report.

Posters

There are two types of posters (also called large-format plots). The first is the summary poster like the kind you see at a conference. The other is the detailed poster that shows all the nitty-gritty attributes for a particular area on a large enough piece of paper to fit them all in. The detailed posters are like architect's plans in that they are not exactly meant for long-range viewing, what with all their minute details laid out in smallish font point sizes. In many ways, both poster types contain the same things: scale bars, data sources, authorship, titles, and so on. In fact, they may vary only in the amount of detail shown in the map element. However, the summary poster can differ from the detailed poster in that it can contain more explanatory text (like methods and results) and fewer annotations on the map element. The detailed poster, on the other hand, does not contain summary text for lay readers and instead contains detailed text and symbology in the style of the profession for which it is made. It is also likely to contain large amounts of annotations and in-depth graphs for its professional audience.

Summary Posters

When it comes to summary posters, there is a lot of information out there on design techniques. It usually consists of little snippets by the organizers of conferences aimed at those who will be submitting posters to an annual gathering. Sometimes these can yield useful and correct guidelines. However, there are many bad suggestions out there as well. For example, I once saw the recommendation that poster titles should be at least three inches tall, or 100 point. But a 100-point font is nowhere near three inches tall! Not only that, but there are entire websites devoted to selling predesigned poster templates, all of which feature tons of superfluous and confusing details like innumerable separation lines and a zillion places to put a zillion logos. So I am obviously not a fan of those. My point is that you need to do due diligence when following someone else's guidelines. Consider the current design fashions as well as your particular data's needs. Do not be afraid to go outside the box if you

The discussion in the "Posters" section focuses on posters that will be printed out one time, or perhaps a handful of times. *Offset printing*, which is the term for printing big runs of large maps, uses the same design principles found in this section. Thankfully, it's much easier today to find a reasonable price for a large print run than at the time of this book's first publication. There are also many new services for print-on-demand that can do large-format prints as they are ordered, as well as handle the customer exchange for you.

have a strong case for doing something different from the norm. Above all, do not get stuck using new technologies to create the same old drab posters. For example, occasionally a poster can still be seen at a conference that looks just like the old-fashioned posters, before the days when people had access to graphics software and large-format printers. These posters have six sections, one for the introduction, one for the methods, and so on. Unless you are trying to go for a retro look, you can do a lot more with a poster now that you don't have to tape together six letter-sized sheets. By the way, titling the sections of your poster with words like *introduction* and *methods* is generic and boring.

What might be a better way to go? Well, you could probably gain a lot of insight into great poster design by studying the design processes of those who know the poster-as-information format the best—architects and landscape architects. Find a book recommended for freshman students in one of those disciplines that goes into the proper ways of conveying information on paper. Or explore the websites devoted to such topics, audit a class, or pick up a copy of one of their trade journals. Any of those activities will boost your design skills immensely if you are printing out large-format posters regularly in your work.

Detailed Posters

The detailed poster, meant to be seen close-up, includes many map layers, complicated symbols, and a lot of annotation. This aids visual analyses such as quickly looking up the location of a drain outfall for a utility customer or determining the types of soils in a potential build site. These are reference maps in the truest sense. That makes it imperative that any and all information that might be needed to adequately assess an issue is included. In this case, if you are in doubt you should err on the side of leaving the data on the map.

Margin elements can be complex and lengthy on these detailed posters. All margin information, usually even the legend and sometimes scale bars, is confined to a neatly organized rectangular space found at the bottom or right-hand side of the poster. The whole works are included as far as margin elements go: date, data paths, data sources, detailed legend information, signatures, title and subtitle, to name a few.

Dual Purposes

I find that clients will often request digital copies of what was originally intended to be a paper map in report or poster form for use in slide presentations or to be put on the client's website. I first determine what, exactly, the digital map is going to be used for and then modify it accordingly. For example, I will take off most of the margin information associated with a large-format poster before making a digital copy for a slide. The margin information could distract from the presentation, and the presenter can simply tell the audience any of the important details they might need to know. Conversely, a letter-size map that you originally made for a report might be modified for a slide presentation by adding detail to it. For example, a letter-size map of the United States might have a lot of leader lines connecting the state names to the smaller states. When you modify it for a slide, however, you could delete the leader lines and place the labels within the polygons since the slide will be much larger than the letter-size map. Obviously, if you are transforming a map from a slide graphic to a letter-size report, you may need to do the reverse: delete detail and move elements around to fit on the new page size.

Endnotes

1. Lawrence Lessig, Professor of Law, Stanford University: http://www.lessig.org/. To see one of his presentations, see http://www.youtube.com/watch?v=mw2z91VW1g (accessed January 29, 2014).
2. A stand-alone 8.5-inch by 11-inch map, such as a handout, does need that supporting information, either in an architect's style box at the bottom or right-hand side of the page (see "Margins" section in Chapter 3) or as text on the flip side of the paper.

Suggestions for Further Reading

Berkun, Scott. *How to Give a Great Ignite Talk*, http://scottberkun.com/2009/how-to-give-a-great-ignite-talk/ (accessed October 17, 2013).

Reynolds, Garr. *Presentation Zen* blog, http://www.presentationzen.com (accessed October 16, 2013).

Samara, Timothy. *Publication Design Workbook: A Real-World Guide to Designing Magazines, Newspapers, and Newsletters.* Gloucester, MA: Rockport Publishers, 2005.

Tufte, E. R. *The Visual Display of Quantitative Information.* 2nd ed. Cheshire, CT: Graphics Press, 2001.

Study Questions

1. What is DPI and how is it calculated?
2. Given a DPI of 300, how many inches would a printed 600-pixel by 800-pixel map be?
3. What steps might you take to modify the map found at this *Washington Post* web page so that it would be suitable for a slide? http://www.washingtonpost.com/wp-srv/metro/daily/graphics/blossomsMap_032505.html. Briefly describe at least two things you would modify.
4. Name two compression-based image file types and two uncompressed file types listed in this chapter.
5. List two problems with using graphics software to manipulate map data. What are the solutions?

Exercise

Find and download an SVG-based icon set meant for maps. Use a graphics program to clip out one of the icons and save it separately. Include in your project files both the unclipped and the clipped SVG files. Extra: use the clipped icon in a map.

8

Projections

Let's be honest, a poll of those who specialize in geospatial analysis would probably indicate that not many—perhaps as few as 20%—could give a hoot about projections. It's not a general ignorance that fuels this apathy; it comes from the fact that the majority of those employed in geo are constantly creating maps and working with data that covers exactly the same geographic region, day in and day out. In these cases, the standard projection for the location has already been decided on—sometimes even legislated—and the geoprofessional therefore works with that projection without questioning it. Sometimes this is a reasonable approach as there are many areas that demand the geoprofessional's attention other than choosing projections, and an assumption that someone else has already thoroughly investigated the issue to choose the best option is not entirely crazy. However, there are times when a much better projection could be used than that standard one, especially if the map purpose is clearly at odds with particular traits of the standard projection (see shaded box below). Furthermore, geoprofessionals who wish to expand into mapping regions that are outside their usual jurisdictions should, or rather must, obtain a basic understanding of projections, and most importantly, their relative strengths and weaknesses.

There is much existing learning material on projections, and some of it is even quite good whether you are looking for a concise explanation or something very thorough. Concise treatments will cover the basic definitions of projection-oriented nomenclature as there are quite a few terms that are specific to this subject and in which it is essential that you become conversant. The more useful of these concise treatments also explore different methods of choosing projections that are suitable given a set of potential variables related to the map's location, size, purpose, and so on. Texts that go into depth on projections also delve into the mathematics that underpin the most common types. This chapter takes the middle ground in going over common definitions, presenting an organized set of considerations for choosing an appropriate projection, and also providing several real-life examples from practicing cartographers. The text avoids the mathematics, but this in no way diminishes its importance, and particularly if you happen to love geometry, you will be well served to seek out more depth in that area from other texts.

Often, you will read that the map's purpose should tell you what projection to use. However, that's only one criterion out of many to consider. The strengths and weaknesses of projections are discussed in this

Dan Bowles is the head of the Cartographic Division of *Australian Geographic* and also the head of a separate company, aptly named The Cartographic Division. He deals with projection challenges in the many maps he is tasked with creating of the Australian continent. The general public is very familiar with the Mercator projection's depiction of the continent, which conveniently displays the state divisions as perpendicular lines, but also badly misrepresents the area within the states. A map of any large area in the Mercator projection vastly distorts area and distance. Using the standard national projection instead—the Geoscience Australia Lambert, a conical projection—mostly ameliorates the area distortion issue (while also helping to preserve true shape), but provides a somewhat less familiar view of the area. Each map that Bowles creates, then needs to consider the pros and cons of each projection before he proceeds (see Figure 8.1).

chapter: (1) in the key distortion considerations of area, angle, direction, and distance; (2) in the key map considerations of purpose, scale, directionality, location, and familiarity; and (3) in the key directional types: cylindrical, conic, and azimuthal. The map purpose informs the distortion compromises to be made and the directionality of the map's location informs the choice of directional type (see Figure 8.2). But before expanding on each of these three concepts, an overview of some common terms is necessary.

To begin with, the terms *coordinate system* and *projection* can be confusing, so it is necessary to define them appropriately. A *coordinate system* can

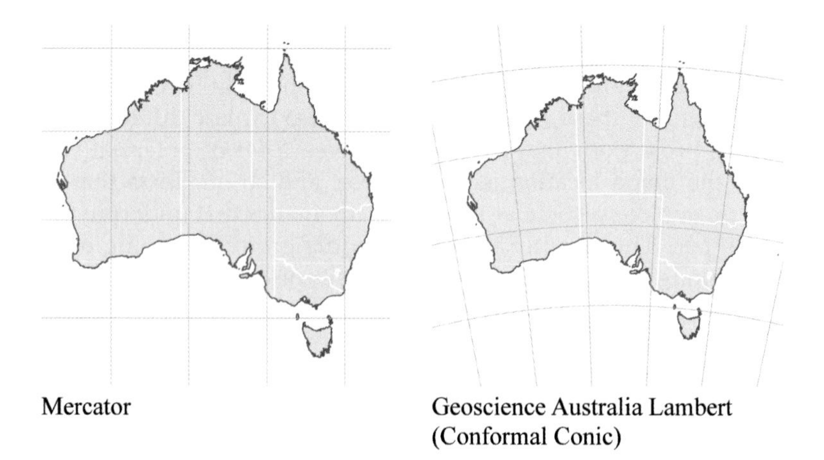

FIGURE 8.1
Australia, projected in the Geoscience Australia Lambert projection and the Mercator projection. Modified from maps by Dan Bowles, The Cartographic Division.

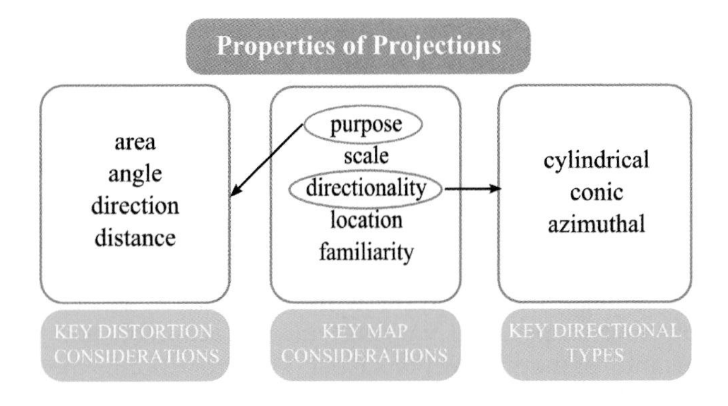

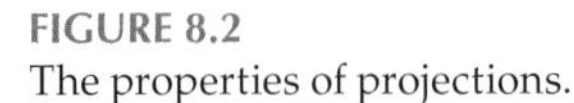
FIGURE 8.2
The properties of projections.

be one of two types: geographic or projected. It will commonly be defined in terms of a datum, spheroid, standard parallel(s), central meridian, and x and y shifts. *Datum* has a few different definitions, but in the most-used geographic information system (GIS) sense it refers to surface positions relative to the center point of the ellipsoid used in the coordinate system. Surface positions among different datums can be as much as hundreds of meters different for the same features. Ellipsoids and spheroids are ways of describing the shape of the Earth where the ellipsoid is a simplistic definition of the major and minor diameters, while the spheroid takes into account the nonregular, "bumpy" surface, and is much more mathematically intense. Common datums include the North American Datum (NAD) 27, North American Datum (NAD) 83, and World Geodetic System (WGS) 84. Differences in datums are only visually noticeable in large-scale maps. The other coordinate system parameters are prechosen for that coordinate system, but remember that all coordinate system parameters can be customized if you wish. The measurement type is also specified in a coordinate system (e.g., decimal degrees in a geographic coordinate system or meters or feet in a projected coordinate system).

A *geographic* coordinate system (GCS) is that system by which lines of latitude and longitude are arranged on the three-dimensional (3D) surface of the Earth. A good way to remember which lines are which is to think of latitude lines as being like the rungs of an upright ladder; they run east–west. The zero point is the longest latitude line, and is called the *equator*, which is the imaginary separator between the Northern and Southern Hemispheres. The longitude lines run north–south, through the poles. The zero point for the longitude lines is most often the imaginary line referred to as the Prime Meridian. It is also called the Greenwich (*Gren-itch*) Meridian since it passes through the Royal Observatory in Greenwich, England. The zero point doesn't have to be at the Greenwich Meridian, but it is what most authorities use. An elevation might also be specified.

A *projected* coordinate system, also known as a map projection, is the mathematical transformation of a geographic coordinate system onto a two-dimensional (2D) surface. In other words, when you need to place real-world features onto a flat page or device and, naturally, scale them down, you need a projection. This is, in fact, the most fundamental definition of a projection and it can extend to anything that you are "projecting" onto a flat surface, such as a movie image or in our case, maps. The math behind projecting the global map image onto a flat surface is both interesting and complicated. Many, many attempts have been made through history to better the math behind projecting maps, with varying success, accuracy, and popularity. In fact, thousands of map projections exist. The projection is the foundation for mapping. Without it, we'd be plotting all our data onto globes.

Arthur H. Robinson, who was a University of Wisconsin geography professor from 1947 to 1980, made one such attempt at creating a projection in the early 1960s when the company Rand McNally asked him to identify a projection that minimizes certain distortions, such as area and distance. Robinson was unable to find a projection that comported with the McNally requirements, so he built his own. He went about it a little differently than other projection pioneers in that he started with an idea of how he wanted it to look and behave, used trial and error with various formulae, and then came up with the overarching formula to make it work. Most projections are created by starting with a mathematical concept first, and then basically accepting how it looks and behaves once it is plotted out. The Robinson Projection wound up striking a nice balance among certain desirable features. One of its great accomplishments was minimizing area and distance distortion in the places where we need it the most: in the regions with the highest populations, which correspond to the Earth's two temperate zones (see Figure 8.3). In addition to the Robinson projection, there are hundreds of other projections. Every one necessarily distorts something—or many things—since transforming the global Earth onto a flat surface is impossible to do with complete accuracy in all measures.

Now that the terms coordinate system and projection are firmly disambiguated in your mind (right, right?!), the next thing anyone learning about projections should do is look at a visual of the most common projections in use today

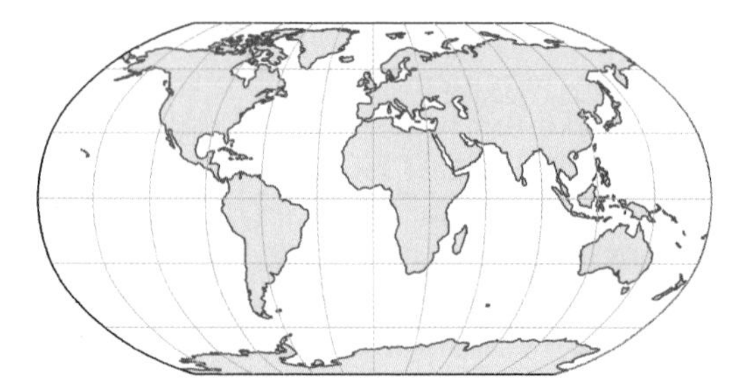

FIGURE 8.3
The Robinson projection.

to literally *see* how they vary. You need to be able to play with the different projections interactively to really get a sense of what's going on. When I was in my first few years as a geoanalyst, a non-geo colleague asked me what projections are. I floundered for an explanation, and the words I did manage to put together didn't seem to make the look of confusion on his face go away. Since we were sitting at my computer, I finally realized that we could simply use my GIS software to project and reproject a world dataset on-the-fly. This, I accomplished in a matter of minutes, and like magic his confusion dissipated. Don't underestimate the power of the visual explanation with regard to projections. To that end, make sure to visit Mike Bostock's Projection Transitions interactive projections tool right now.[1] For starters, take a look at Mercator (used in most digital maps as of this printing), the Winkel Tripel (used by *National Geographic*), and the Transverse Mercator (used by many national agencies).

Distortions: Where Projections Fall Flat

The interactive projection visualization that you've just looked at immediately makes it apparent that projections can be good at certain things and bad at others, and never good at all things at once. More specifically, projections can exhibit distortion in *area*, *angle*, *direction*, and/or *distance* (remember these as AADD). Some projections preserve one or more of these properties and some compromise on all four of the properties. Now that you've seen, visually, how these properties differ, it's important to get a finer-grained understanding of exactly how they differ, which is covered in this section. Once that's accomplished, remember that the final choice of projection will usually require that you investigate more in order to definitively quantify the suitable candidates' distortions in terms of error. For example, a goal of minimizing area distortion is a good first start, but once that goal is established and a suitable list of projections is produced that minimize area distortion, you must then decide from there which one will be most adequate depending on a variety of other factors.

The *area* of objects on the map, most notably the continents in the previously mentioned interactive visual, varies from projection to projection and also sometimes within a map of a single projection. Some projections preserve area throughout the map, while others distort all or part of the map areas; this is particularly noticeable in the Mercator projection, where the poles are shown as much larger than they truly are relative to areas near the equator. Projections that preserve area are called *equivalent* or *equal-area* projections. These are the projections you want to use when showing how a particular dataset is spatially distributed such as population, climate, or food consumption. Even so, you must be cognizant of any limitations that might make one equivalent projection better than another. For example, the Gall-Peters projection is an equivalent projection, but it distorts the shape of the landmasses to such a great extent that it has limited utility. The sinusoidal projection is one of the earliest equivalent projections, and is in use today as the predominant projection for National Aeronautics and Space Administration Moderate Resolution Imaging Spectroradiometer (NASA MODIS) satellite data.

Second, *angle* (shape) is distorted to varying degrees on many projections. Projections that preserve angles are called *conformal*, and are characterized by maintaining a 90-degree angle between perpendicular graticule lines. Conformal projections are suitable for large-scale maps where local navigation is required, but are not suitable for medium- to small-scale mapping, where, indeed, they cannot maintain a constant angle. In general, map scales of 1:100,000 or larger are appropriate. At any scale, angle can't be preserved at the same time as area, and in fact, when maintaining a constant angle, the area is usually much distorted. Therefore, a conformal map is not appropriate for any mapping where you need to compare data based on relative size. Suitable uses for conformal projections include topographic maps, wind direction maps, and certain types of navigational maps such as military navigation. The Mercator projection is conformal as is the Lambert Conformal Conic projection—which is used by pilots for plotting a true course between two points.

Digital mapping is currently dominated by the Mercator projection, which has met with some derision from the mapping community ever since non-mappers adopted that projection as standard for web mapping and it became ubiquitous. The Mercator projection is especially not ideal for most kinds of data displayed on a global scale, particularly when that data has anything to do with area. For example, a worldwide map showing protected areas like national parks and their equivalents, if produced in a Mercator projection, will make Canada and other areas far north of the equator appear to have vastly more protected area in comparison with those near the equator. However, until very recently, digital mappers didn't have much choice but to display their interactive map data in the Mercator projection since that is the only choice for most web-mapping application programming interfaces (APIs). This may or may not be changing as other developers come into the fray with new APIs that allow for the choice among a multitude of projections.

The third type of projection is referred to as *azimuthal* and it preserves *direction*. In an azimuthal projection, a world map is circular, not rectangular, and the point at which the map would "touch" the Earth (the point that truly matches between the globe and the flat map) is at its center. This means that the direction from the center of the map to any other point on the map is correct. This is somewhat confusing because, while we are talking about angles here, azimuthal projections don't preserve all angles like conformal projections, and the angles aren't 90 degrees. In an azimuthal projection, the angle between the center of the map to a point (a) and the center of the map to another point (b) is true. Interestingly, a few projections are considered two-point projections, where the angle is preserved from those two points to any other points instead of just one. Shape and area are very much distorted far away from the center point (or two center points), though in some cases this distortion is acceptable for datasets (such as bird migrations) that may require it. These projections aren't common in thematic types of maps, but should definitely be considered more, especially if a new "perspective" is sought, or if any directional data is to be displayed, especially when the data spans a hemisphere or crosses a pole. Gnomonic, stereographic, perspective, and orthographic projections are all azimuthal projections.

Finally, projections that preserve *distances* (scale) are called *equidistant*. Similar to azimuthal projections, where the direction is preserved between the center point (or, less commonly, from two points) and the other points on the map, equidistant projections preserve distance from one point to all others, but not among all of them. Any map that shows linear distance data, such as the travel distances for goods sold, would be particularly suitable for an equidistant projection. One of the first projections ever created is equidistant—the Plate Carrée. Sinusoidal, Werner cordiform, and equidistant conic are some other equidistant projections.

One tool that's been developed to visualize the errors in these factors in a quantified and visual way is the Tissot's ellipse. With this tool, angle and area distortions are computed for a projection, and ellipses are plotted on a globe at regular intervals. Each ellipse shows you how distorted the projection is at that point on which the ellipse is plotted. If the ellipses are different sizes, then area is distorted proportionate with the area change. If the ellipses are not perfect circles, then the angles are distorted proportionate with the angle change. In this way, you can get a quick understanding of how much distortion there is and also where that distortion is. A perfect circle indicates no distortion. The Tissot's ellipse technique is a good thing to be familiar with—you may hear other mapmakers talk about Tissot's ellipse, also called Tissot's indicatrix—but with the interactive visualization tools we have now it's not as essential to our profession. For one alternate way of looking at projection distortion levels in a quantified manner, see Syntagmatic's Comparing Map Projections.[2]

Map Considerations: Choosing a Suitable Projection

Now that we've seen how projections can differ in distortion, we can begin to see that those distortions are going to inform our choices when picking a projection to use for any particular map. In fact, these distortion classes (also called *preservation classes* if you want to speak more optimistically) are important when considering the map purpose, which is the first item we'll discuss in this section on choosing projections. If you need to pilot a plane, if you need to plot population distribution, or if you need to track bird migration patterns, you can look at the distortion classes to pick a suitable projection based on that purpose. In addition to map purpose, however, there are also a few other considerations of which you should be aware; these are scale, directionality, location, and familiarity. If I were to pick two to focus on first, it'd be purpose and scale. Distortion becomes less of an issue with larger map scales, since there is just less potential for distortion across small areas. Large-scale maps, therefore, don't need as much emphasis placed on distortion when choosing a projection. This is a good thing to know right at the beginning. Do be mindful, however, that even though large-scale maps don't need as much thought when it comes to distortion, they do need thought when it comes to which *datum* you are going to use, since at large scales small differences in datums become a large problem (see earlier datum discussion).

The purpose of the map will indicate which of the four distortions you can afford to compromise on and which you can't—area, angle, direction, or distance. For example, maps of population distribution should be created with an equal-area projection in order for the map reader to accurately compare population ratios. Most local road maps should be in a conformal projection. Medium- to small-scale maps showing the direction of a phenomenon such as flight paths can be most easily understood in an azimuthal projection.

For most maps, once the map purpose is identified and a set of acceptable projections is chosen based on minimizing distortion, the next thing to do is think about the scale of the map. *Scale* here means whether the mapped features will be small, medium, or large. So let's say you are creating a navigational map. This map purpose has led you to create a list of suitable conformal projections, since you are interested in angles being preserved across the entirety of the map. If the scale of the navigational map will be large (a portion of a town, for instance), then the conformal projections will be appropriate. If the scale is medium or small (the western United States, for instance), then a conformal projection will not be appropriate and alternative types of projections need to be considered. In this case, you would go back to the various distortion classes and take another look at whether angles need to be accurate or whether you can compromise on those due to the scale. Medium- and small-scale navigational maps aren't common and may require navigational disclaimers explaining their error potential.

Of course, in reality, the scale of the map and the distortions that you are willing to compromise on will be considered more or less simultaneously rather than stepwise. It is just more convenient for an instructional text such as this to explain them sequentially. Finally, when looking at scale in particular, the benefit of medium or small scales is that you can be pickier about the aesthetics of a projection. A few of the more outrageous examples involving aesthetics might be: you can look at the projections in terms of whether the continents look like hung laundry (Gall-Peters projection) or whether Greenland is the size of Africa (Mercator projection) or whether its butterfly resemblance is too distracting (Waterman projection).

For large-scale mapping, in particular, we need to turn our attention to some common terms that are important: ellipsoid, geoid, and datum. An *ellipsoid* is a spherical shape where the center is larger than the top and bottom. In map terms, an ellipsoid is an accurate representation of the Earth's true shape in terms of the Earth having a somewhat larger radius at the equator and smaller radii extending to each of the poles. It is also called an *oblate spheroid.* A *geoid* is the same thing, except a geoid is a "better" representation of the variations in the peaks and valleys of the Earth that are caused by the Earth's gravitational field, whereas an ellipsoid characterizes a continuous surface. The geoid, in other words, is more accurate and is based on sophisticated satellite data as opposed to the mathematically defined ellipsoid model. Ellipsoids are computationally less intensive than spheroids. *Datums* are created from ellipsoids and are a mathematical way of defining locations by using a point where an ellipsoid matches with the geoid. A datum is a way of describing which spheroid you're using as well as the point at which that spheroid coincides with the real surface of the Earth on your map. A datum is important for large-scale cartography, such as transforming in-field measurements into a map of a town's invasive plant species, where the precise locations and scales need to be preserved. For general maps of data distribution, a datum is not as important. The WGS 84 is one of the most commonly used datums now.

After map purpose and scale, the directionality of the map is the next item to examine. The directionality of a map is the tendency of the underlying features to be oriented mostly east–west or north–south. Interestingly, quite a few mapped features run east–west. In fact, Jared Diamond, professor of geography at the University of California, Los Angeles, contends that climate, soils, and biomes are all east–west oriented, and more recently researchers have discovered that historical empires were also east–west oriented.[3] Eurasia, the United States, Australia, and many other landmasses or pieces of landmasses are also oriented east–west, though the Americas and Africa are north–south oriented. For east–west landmasses, suitable projections include the Albers Equal Area Conic and Lambert Conformal Conic, as long as the map doesn't encompass an extent larger than a hemisphere. To determine the orientation of your features, take a look at their shape on a globe, not on a map that's already been projected. See also the next section on directional types.

The next item on our list of considerations is the location of the data. Is the majority of the map going to be in an equatorial region, polar region, or somewhere between? In other words, what is the latitude of the center of your mapped features? Cylindrical projections are well suited for equatorial maps, azimuthal projections are great for polar maps, and conic projections are useful in the areas between the equator and the poles. Be aware that it is possible to use your geo software to customize the parameters, such as the meridian and x-y shifts, in order create a projection that is exactly suited to that particular locality. See also the next section on directional types where cylindrical, azimuthal, and conic projections are discussed in more detail.

Next, before becoming enamored with a projection, you should also be aware of the common projections used for your location of interest. If you work for a US county, the standard is usually a State Plane Coordinate System that is fine-tuned for that part of the country. If you are working on a generalist map of a bigger area, familiarity also comes into play when working with the shape of the features. Spain, for example, looks abnormally squished in the east–west direction in a sinusoidal projection. A sinusoidal projection is great for performing analyses on geographic data that span the world where preservation of area is important, but you wouldn't use a sinusoidal projection for a general map of Spanish tourist sites since it distorts shape a great deal. Know the written and codified standards for large-scale jurisdiction mapping and also know the most familiar shapes for a given mapped area.

Directional Types

As discussed earlier, the predominant directionality of the map's main features influences which of the following projection directional types to choose: *cylindrical*, *conic*, or *azimuthal*. In simple terms, a cylindrical projection is best for features that extend north–south. A conic projection is best for features that extend east–west. And an azimuthal projection is best for features that have more or less equal north–south and east–west directionality. These three projection types are best described visually (see Figure 8.4).

The cylindrical projection creates a rectangular map at the world scale where the lines of latitude are at right angles to one another, resulting in a regular grid pattern. However, the distances between the longitudes widen near the northernmost and southernmost portions of the map. Cylindrical projections are very common, and the lay map reader is familiar with them—a good argument for using them. The Mercator projection is the most well known of this type. A useful variation on the cylindrical projection is the pseudocylindrical projection, where instead of a rectangular map, the map

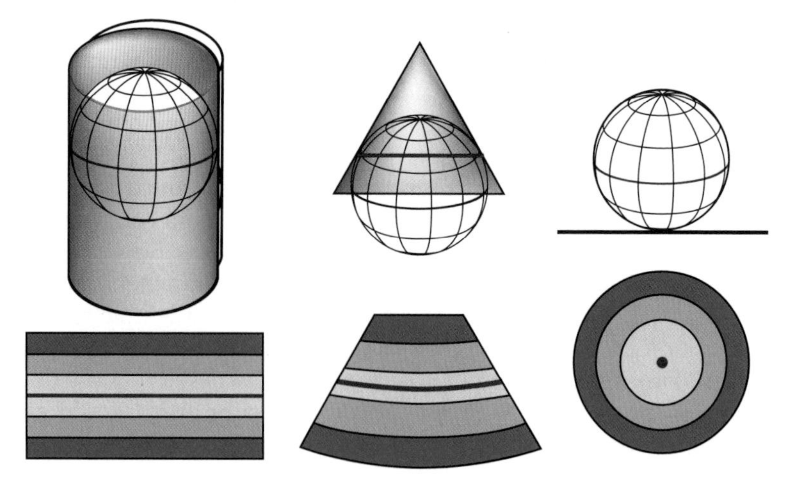

FIGURE 8.4
The projection planes of cylindrical, conical, and azimuthal projection classes.

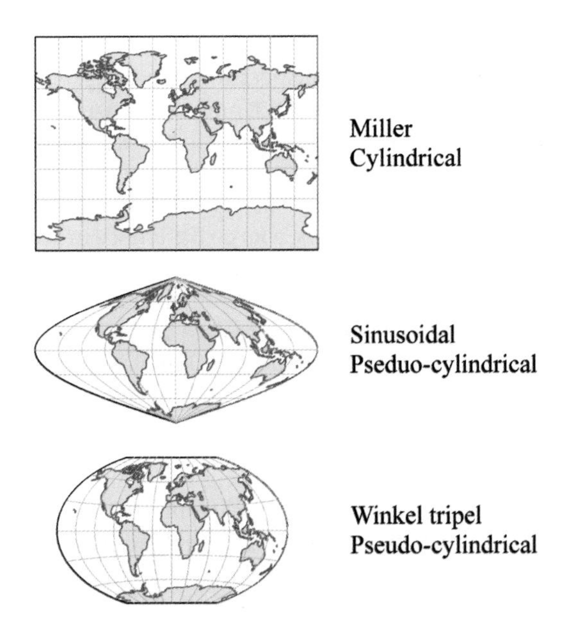

FIGURE 8.5
Examples of a cylindrical projection and two pseudocylindrical projections at the world scale.

looks a bit like a UFO at the world scale, with the poles coming together as points, which ameliorates to some extent the distance distortion in those locations. Some are less pointy and appear more like flattened pumpkins with the poles somewhat converging but not all the way. The Winkel tripel (used by *National Geographic*) is a pseudocylindrical projection (see Figure 8.5).

Conic projections look like portions of circles (pie slices, perhaps) at the world scale. In these projections, the latitude lines are circular and the lines of longitude, while straight, are not parallel to one another and radiate out from the center so that the angle between the latitude and longitude increases as they reach the outer edges. These projections are great for mapped features that are mostly in the middle of a hemisphere, where their distortions are smallest, as opposed to equatorial or polar locations, where their distortions are greatest. (Incidentally, when you look at a typical conic projection you, like me, might realize that the shape lends itself particularly to a lampshade. Apparently, not many people are like me, because as of this printing, there aren't any commercially available but there are rumors that projection enthusiasts have constructed them from time to time.) Tennessee's state plane coordinate system is a Lambert conformal conic projection, which is fitting due to Tennessee's east–west orientation.

Azimuthal projections were described a bit earlier. They are sometimes referred to as *planar*, and they look like full circles at the world scale instead of the conic projections' partial circle appearance. They have circular lines of latitude and straight lines of longitude that radiate outward, also resulting in

> The map projection impacts other aspects of map design in several ways. For one, spherical projections should not have north arrows, since north is not in a constant location. If direction needs to be indicated on a spherical projection, it can be visualized with a graticule grid. Similarly, projections that have varying scale across the map should not have scale bars on small-scale maps where there will be a large difference in scale depending on where you are measuring. If a scale is absolutely necessary on a small-scale map, the scale should be shown as a variable that changes depending on the latitude. For example, for the Mercator projection, one kilometer at the equator will be much smaller than one kilometer at a pole, so this can be indicated with a graph of latitudes and their respective scales.

larger angles between the latitude and longitude lines, just as with conic projections. All directions (i.e., azimuths) from the center point are preserved, but distances and shapes become increasingly problematic as you move away from the center. Azimuthal projections can provide a unique perspective on the mapped features that no other projections can; they can emphasize the continuity of the continents and the flow of features around the globe, especially if they all radiate from a single point.

Choosing a Projection in the Real World

Leaving the theory aside, we'll take a look at a few examples of how and why professional cartographers decide on projections for specific maps. These are mapmakers who have had to consider all of the information thus far presented as well as weigh the needs of their particular map and audience. While the eventual choice of projection is subjective, I think you'll find that every mapmaker takes these decisions seriously and makes a choice with much deliberation.

Evan Centanni, of Political Geography Now (http://polgeonow.com), was asked to make a map of estimated missile systems ranges using available public data (see Figure 8.6). The most important distortion to avoid was distance from the missile launch points since range distance was the focus of the map. While users aren't expected to measure actual distances on the map, it was still important to choose an equidistant projection so that the distance bands would appear as perfect circles. It should be noted that the range buffers were also calculated in an equidistant projection. Also, because the scale of the map is on the smaller side, distortion in distance would need to be appropriately handled via an equidistant projection. To that end, Centanni decided on an azimuthal equidistant projection centered in northern Taiwan. Due to the purpose of the map, distortion in area and shape weren't concerns, but Centanni did still consider them in the context of user familiarity. Ultimately, he decided that the equidistant projection he chose was adequate

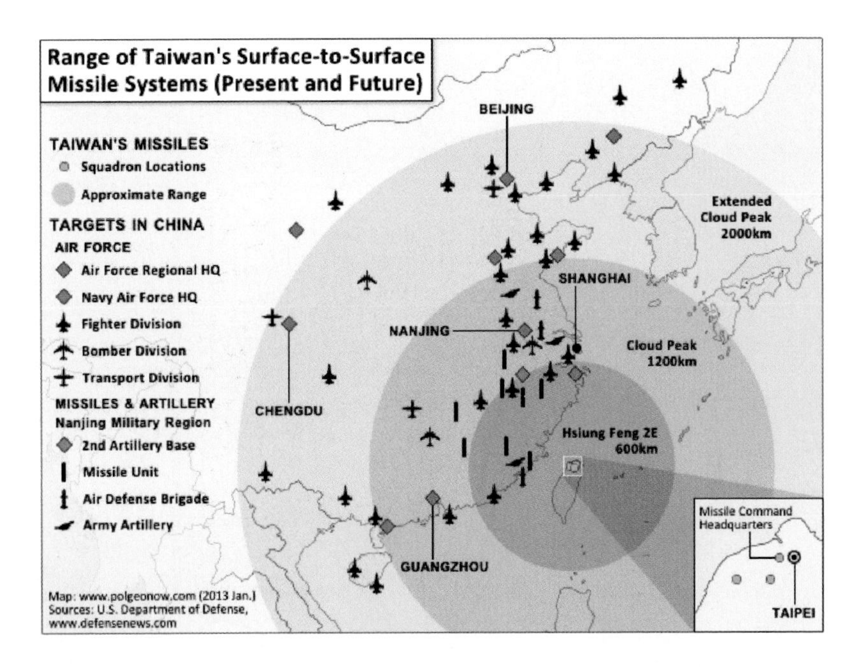

FIGURE 8.6
Range of Taiwan's Surface-to-Surface Missile Systems (Present and Future) by Evan Centanni, http://www.polgeonow.com.

in preserving area and shape enough to be recognizable to map users. While Asia is a bit tipped to the right compared to most world map projections, it is barely enough to notice, and though this projection distorts Africa and the Americas greatly, they aren't shown in this map.

Hans van der Maarel, of Red Geographics, produced a map of English and Norwegian activities in the Antarctic in the 1950s, 1960s, and 1970s (see Figure 8.7). It also shows the political situation of the late 1950s, hence the title. Van der Maarel used a polar stereographic projection for this map due to its polar location. However, it's rotated by 60 degrees because (1) the page size was limited and (2) this configuration shows the three overlapping claims (UK, Chile, and Argentina) at the 12 o'clock position, giving enough room to place the labels. It is common to change the orientation of a polar map in this way, but be forewarned that it will cause the map user to need extra time for interpretation. This map was created for a print book publication, which lends itself more to extra perusal time by the reader than, say, a simple website would.

This chapter is not meant to be an exhaustive primer on projections. Indeed, the mathematics behind projections and the varying tools for examining their properties are not detailed here. These are both worthy subjects for additional study, but go beyond the scope of this cartography-minded text. You are advised to learn more about projections as you go about your studies. With that end in mind, take a look at the list of resources that follows to gain more depth in this important subject.

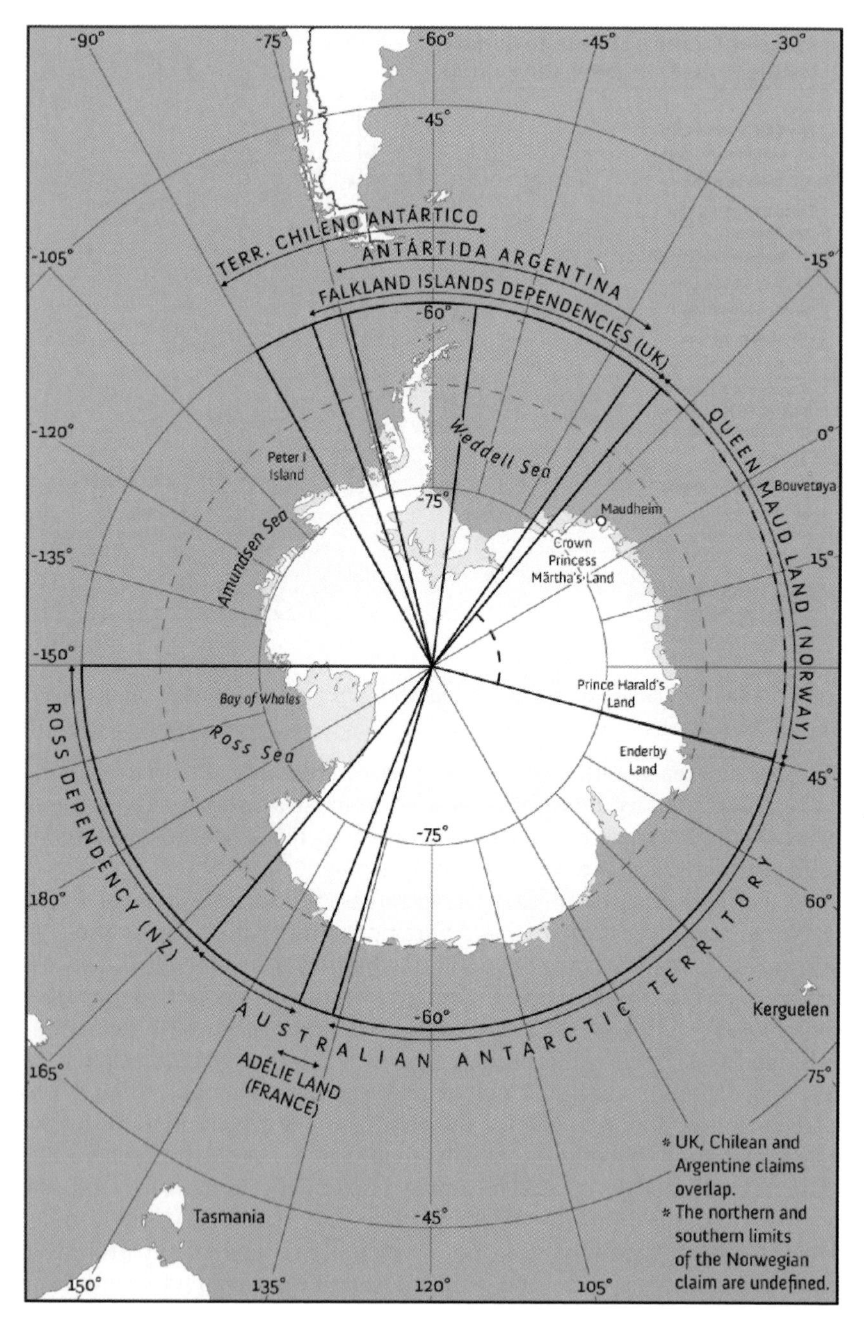

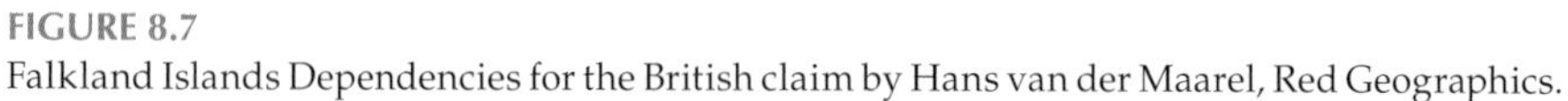
FIGURE 8.7
Falkland Islands Dependencies for the British claim by Hans van der Maarel, Red Geographics.

Endnotes

1. Projection Transitions: An animated projection demonstration in D3, by Mike Bostock, suitable as a reference and learning tool for mapmakers irrespective of platform, http://bl.ocks.org/mbostock/3711652 (accessed September 23, 2013).
2. Comparing Map Projections: An interactive projection demonstration in D3 with 18 projections illustrated along with their respective distortion levels in overall projection distortion (acceptance index or ACC), scale, area, and angle, http://bl.ocks.org/syntagmatic/3711245 (accessed September 23, 2013).
3. J. Diamond *Guns, Gems, and Steel: The Fates of Human Societies*. New York, Ny: W. W. Norton & Company, 2005; P. Turchin, Adams J. M., and T. D. Hall, "East–West Orientation of Historical Empires and Modern States." *Journal of World-Systems Research* (December 2006): 219–229, http://www.jwsr.org/wp-content/uploads/2013/03/jwsr-v12n2-tah.pdf (accessed September 23, 2013).

Resources

Two in-depth resources for projection information:

Snyder, J. P. *Map Projections: A Working Manual*. 1987. US Geological Survey (USGS) Professional Paper 1395.

Bugayevskiy, L., and J. P. Snyder. *Map Projections: A Reference Manual*. London: Taylor & Francis, 1995.

Two additional resources for projection information:

Xkcd, a web comic, has an amusing take on what your favorite map projection says about you. It might be solely responsible for making *dymaxion* and *waterman butterfly* household names (or at least better known). See http://xkcd.com/977/.

Methods for choosing projections for small-scale maps are included in Canters, F. *Small-Scale Map Projection Design*. New York: Taylor & Francis, 2007.

Study Questions

1. Briefly discuss the differences among the three directional types and what you must consider when selecting a directional type.
2. Select a projection for a locality that you know and describe why you selected it.
3. What is the difference between an ellipsoid and a spheroid?

4. At what latitude and longitude are the equator and Prime Meridian located, respectively, in most coordinate systems?
5. Look up an example of Tissot's ellipse (i.e., Tissot's Indicatrix) and describe what it shows for a projection of your choice in terms of the distortions that projection has and where those distortions are.
6. Discuss what projection considerations you would want to investigate for large-scale maps and why, with respect to distortions and datums.
7. Would a map focused on the Mediterranean Sea be best with a cylindrical, conic, or azimuthal projection? Discuss your choice.
8. Pick a real-life municipality, such as a city or county, that uses a standard coordinate system for all or most of its government mapping. List the name of the municipality and its standard coordinate system parameters.
9. What is the difference between a geographic coordinate system and a projected coordinate system?
10. What is the term used for projections that preserve angles (shapes)?

Exercises

1. Create three maps of the Middle East in three different projections: one with a cylindrical projection, one with a conic projection, and one with an azimuthal projection. The map itself should be simple: major seas and gulfs in a single color, country boundaries, a single color for the land, country labels, a title, data source citation(s), and author name. Bonus: include graticule lines.
2. Obtain a local point dataset, such as well points, stop signs, manholes, and so on, and display them on a map in their native datum. Reproject the dataset into a different datum and display the new dataset on the same map but in a different color. The map should have a title indicating the dataset name. It should also have a data source citation, author name, and a simple key with the dot colors and their respective datums. Include a scale bar. The map should have a large enough scale to depict the differences in datums between the two datasets.

9

Zoom-Level Design

> Easier, simpler, and friendlier is the way to go. Web maps should be like toasters: no need to log in, read a manual, or think too hard. Just put in the bread or address and push down on the lever or the mouse button!
> Martin von Wyss, a comment on the Fuzzy Tolerance blog

Whether you currently design digital, interactive map products or are about to—and yes, everyone who isn't already is *about to*—it can save a great deal of time to learn the techniques, constraints, and design skills needed for these types of maps ahead of time. Unlike other mapping concepts like projections, color, fonts, layout, and so on, digital map design is in its nascent stage, and as such there is little established, published information on the topic. There are a lot of great digital maps in existence, but many of the developers, designers, and geoprofessionals who made them did so by much trial and error, with very little formal learning, if only because those learning materials just weren't available. This chapter aims to fill that void and is meant to focus on the key *design* aspects of multiscale digital mapping, which are the multifaceted, unique design considerations of making essentially a different (yet similar!) map for each and every zoom level, with potentially up to 20 zoom levels or more. No problem!

A word of warning: there is a small amount of code in this chapter. It's in a format similar to CartoCSS. Most digital maps today are created with code, whether it's JavaScript, CartoCSS, or something else. The code is there to illustrate a few points, but don't get caught up too much in how things are done in the code, as it will be different depending on what tools you use to create your digital maps. Speaking of tools, there will be different design decisions, constraints, and opportunities available with each tool or set of tools. D3, a JavaScript library, provides the ability to make maps in any projection, while most other technologies don't. TileMill has easy-to-use compositing capabilities for adding advanced artistic-like touches to a map. The Google API has helpful documentation and tutorials on deploying to Android devices, but doesn't allow for custom basemap making. Choose your tools by considering the pluses and minuses of each. A *map stack* is a suite of technologies that are combined to manage data, and make and serve web maps. As of this printing, some of the most popular tools are the aforementioned as well as Leaflet, OpenLayers, and ArcGIS Online.

Why is this chapter about zoom-level design instead of just digital map design in general? It's necessary to make this distinction because the rest of this book has as much to do with the design of digital, interactive maps as

Cartography today is as much about development (i.e., programming) as it is about design and data analysis skills (see Figure 9.1). It's essential to at least know how write a script to automate workflow. Those professionals coming from the sciences, geographic information system (GIS), and cartography camps need to increase their software development skills. Learn HTML 5, JavaScript, and Python to start. Take a free online course, watch tutorials, download some data and build a digital map, and learn the basics of CSS.[1]

A recent job ad for a cartographer listed the requirements for the perfect candidate. In the bulleted list were six items pertaining to software development knowledge such as processes and specific technologies. One item pertained to design. Not a single item pertained to an applicant's ability to manipulate data in meaningful ways. Unfortunately, the lack of emphasis on data and design probably stems from the fact that the firm is primarily made up of software developers. Naturally they are going to get specific about those things which they know the most about, like particular integrated development environments (IDEs), programming languages, and so on, and not so specific about design and data, about which they know less. This shows us two things: (1) cartographers need developer skills and (2) developers who are hiring cartographers shouldn't forget that data and design skills are also very important. In fact, in this particular case, it seems likely that the company really needs someone who is quite strong in data and design, since these are obviously the skills that are lacking in the company. If they can get someone who is strong on all three counts then they need to throw a party, because there really aren't that many candidates, currently, who have expert level knowledge of all three areas.

it does with printed maps. Good design, as mentioned in other parts of the book, is good design no matter where it is deployed—on a piece of paper or a digital device. Colors can be garish, fonts can clash, projections can mislead, layouts can be too cluttered, and data symbolization can be confusing on both paper and digital devices. Those topics are all very much relevant to all maps you make. *Zoom-level* design is a topic, however, that really begs for individual treatment, since it involves things like special scaling considerations, extra data columns, incremental styling, and special testing techniques. This basis from which to start out on your own digital map design journey is sorely needed.

Most digital maps are displayed in the Web Mercator projection. This became the default standard for web mapping when the major online map services adopted it, like Google and OpenStreetMap. In some ways, it seems entirely inappropriate because Web Mercator greatly distorts the area of the continents. Antarctica and Greenland, especially, appear to be huge due to their locations near the poles, where the maps in Web Mercator are

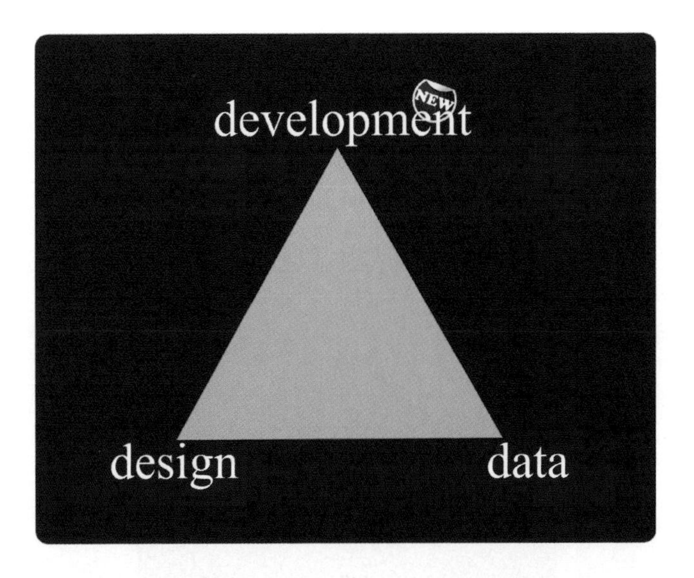

FIGURE 9.1
Cartographers now need skills in three areas: development, design, and data.

stretched the most. However, Web Mercator does preserve direction and shape. In maps that were initially meant to be used for navigation—and indeed this is still the dominant use of online mapping—this was not an entirely poor choice. In addition, Web Mercator is cut off at approximately 85 degrees north and south, which allows the map to be represented as a square, thus aiding in tiling (discussed later). A thematic map, for which area preservation is very important, should not be displayed in a Web Mercator projection, especially at small scales. However, this projection is usually the map developer's only option, even for thematic maps, depending on what technology is being used.

With that requisite discussion on Web Mercator out of the way, let's get into what exactly is meant by *zoom-level design*. Basically, it means that there are special design considerations to take into account with digital maps because they get more and more detailed as the user zooms in and less and less detailed as the user zooms out. The key point is that the design of such multileveled maps is complicated and unique. The lowest zoom levels show continents, zoom 6 shows full views of medium-sized countries, and by the time you zoom in to level 16, you're displaying a city block or two (see Figure 9.2 for some other zoom-level examples). A *zoom level*, then, is essentially the scale of the finished map. It gets complicated though, because the scale isn't necessarily the same type of scale that we customarily use in static maps. And it doesn't stay constant between devices (more about this in the next section).

Another good term to learn is *render*, which refers to the finished map at one zoom level. Render can also be used as a verb, for example, "my software

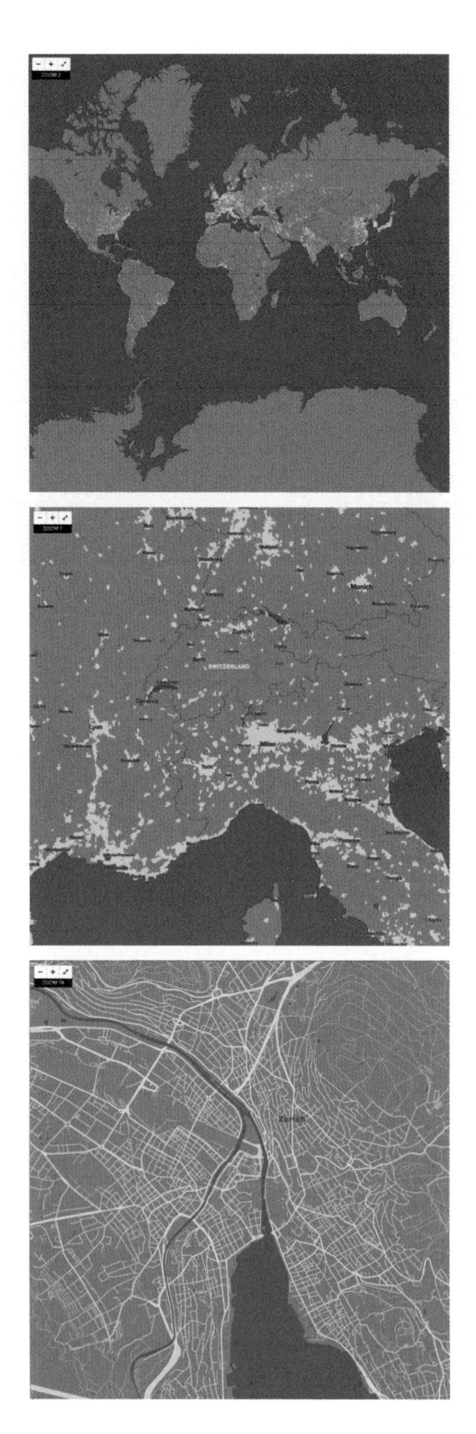

FIGURE 9.2
An example of a digital map at three zoom levels: zoom 2, zoom 7, and zoom 14.

FIGURE 9.3
This is one tile at zoom-level zero.

is rendering zoom-level four right now." Digital maps are created in square-sized pieces that are called *tiles*. A tile is usually, but not always, 256 by 256 pixels in size. At zoom-level zero, most digital maps display the whole world in a single tile (see Figure 9.3). *Tileset* is the term used to refer to all the tiles that make up an entire multiscale map. Occasionally, the term *slippy map* is used to denote any digital interactive map. The term comes from the action involved in panning around a multiscale map, wherein the tiles load on the edge of the page as the panning action takes place.

To be clear, technically the term *digital interactive map* could also refer to a map that is a static image with zoom and pan capabilities, that doesn't increase or decrease in detail at different zooms, but this isn't the focus in this chapter. Furthermore, user interactivity with data is also a major defining factor of digital interactive maps. Designing the interface for such data interactivity is much written about elsewhere in more general user-interface texts.

Multiscale maps necessitate styling changes at each zoom level, or at groups of zoom levels. Things like colors and line widths could be tweaked at each level. Generalized datasets could be swapped out for higher-resolution datasets at higher zoom levels. Label density might need to be increased or decreased. And so on. To accomplish this, you may need sophisticated and well-organized zoom-specific styling rules. How you express these rules will depend on the platform you choose. Researching the various coding techniques for the platform you choose will be important, but is outside of the scope of this chapter. This zoom-level design exploration will start with a general discussion of how zoom levels relate to pixel sizes and scales, a topic that is not well understood even in the established geospatial community.

Zoom Levels and Scales

It is natural to want to match zoom levels with standard paper map scales given that we have standard zoom levels (based on powers of 2) across all digital mapping platforms (0–19) and standard map scales for printed maps. Being able to do so would mean having a good idea of which datasets would be best displayed at which zoom level, depending on their level of generalization. And yes, it is possible to make a very rough comparison. For example, datasets with maximum resolutions of 1:10,000,000 are appropriate for zoom levels 0 through 6. And while that is a general truth, it is not an absolute. These are the reasons why: (1) the pixels per inch (PPI) values on different devices mean that the map scale will be different depending on the device and (2) most web maps use the Web Mercator projection, which means that the scale is variable depending on the latitude.

With most digital maps, the mapmaker doesn't know the device on which the map will be deployed. It could be a computer monitor, in which case the PPI will vary depending on the size of the monitor and the resolution of the monitor. Modern monitors normally have PPI counts in the 90s, but some high-resolution displays can go above 200 PPI. E-readers vary from 150 to 265 PPI, and the newest smart phones can reach the 400s. These are considerable differences. A map viewed on a device with 100 PPI versus a map viewed on a device with 400 PPI is shown in Figure 9.4. To add even more fun to this, we also need to talk about dots per inch (DPI).

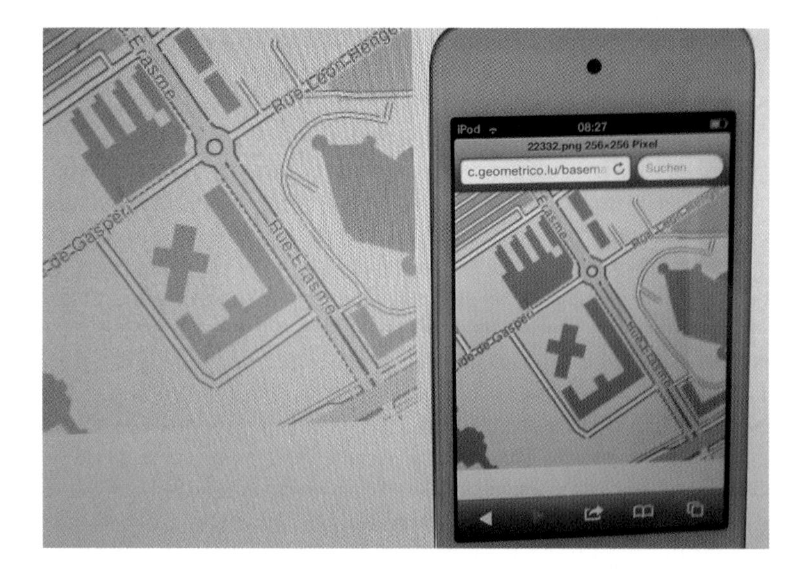

FIGURE 9.4
The same map on two different devices appears to have two different scales.

A typical digital map is exported to tiles at 96 DPI, but sometimes they are exported at other DPI resolutions. This also affects the scale. Charts that equate zoom levels with map scales often assume a DPI of 96. Again, depending on the screen resolution—or PPI—this standard DPI of 96 will appear differently.

The second issue is that of projection. Assuming export to the Web Mercator projection, the scale will vary as much as a factor of approximately 6 between the equator and 80 degrees latitude north and south. Even within the assumed "usable" area of between 70 degrees latitude north and 70 degrees latitude south for the Web Mercator projection, the scale varies by approximately a factor of three. So while a given dataset with a maximum resolution of 1:10,000,000 may be appropriate *at the equator* at zoom levels 0 through 6, it may only be appropriate at the higher latitudes at zooms 0 through 3, for example. Using the equator as a guideline for map scale is thus a convenient, but very imperfect, way of determining dataset usability.

It is therefore recommended that zoom level–to–map scale conversion tables be employed with caution and only as a very general guideline. Always scrutinize datasets at all zoom levels and latitudes to determine best use (see Figure 9.5).

Zoom Level	Approximate Scale
0	1:500m
1	1:250m
2	1:150m
3	1:70m
4	1:35m
5	1:15m
6	1:10m
7	1:4m
8	1:2m
9	1:1m
10	1:500,000
11	1:250,000
12	1:150,000
13	1:70,000
14	1:35,000
15	1:15,000
16	1:8,000
17	1:4,000
18	1:2,000
19	1:1,000

FIGURE 9.5
Approximate map scales by zoom level, Mercator projection, with an assumed 96 DPI. These are rounded and meant to be used for general comparison only. Always be mindful of DPI, PPI, and projection before deciding whether a dataset is appropriate for a zoom level.

Zoom Fields

A zoom field is a dedicated column in the spatial database that indicates the minimum zoom level at which each feature should be shown. The workings of this zoom field are best described by example. Say you have a dataset of worldwide cities. You want to show points for only the largest cities at zoom levels 0 through 5. At zoom 6 you want to start showing the points for the medium-sized cities along with the larger cities, but only up to zoom 10. After zoom 10 you don't want to show any city points, but you do want to show city labels for all sizes of cities including small cities. To create a zoom field for this simple example, create a column in the dataset and call it something like minZoom or scaleRank. MinZoom is preferred as it is a bit more descriptive for those who are unfamiliar with this kind of data field. Populate the minZoom field with the number "0" for all cities with a city population of greater than one million. Then, for the medium cities with populations between, say, 300,000 and one million, populate the minZoom field with the number "6". All the cities that haven't had a number assigned yet, that is, those with populations less than 300,000, should then be assigned a "10" in the minZoom field.

To show how this field is used, we're going to use some pseudocode that is loosely based on CartoCSS. To use the minZoom field in the code that you write to make this map happen, write something like:

```
[zoom > = 0][zoom < = 5][minZoom = 0] {
 point styling parameters such as size and color for the
   largest cities
}
[zoom > = 6][zoom < = 9][minZoom< = 6] {
 point styling parameters such as size and color for the
   medium and large cities
}
[zoom > = 10][zoom < = 19][minZoom< = 10] {
 label styling parameters such as size and font for cities of
   all sizes
}
```

By adding the minZoom constraint next to the zoom level specifications, you are telling the software to only show those features for those zoom levels. By doing this, you avoid having to separate the data into different datasets based on zoom level. Instead, all the city data can be in the same dataset and simply displayed differently based on each city's zoom field number. Another thing that this method avoids is duplication of rows in the dataset. There is no need to have three sets of the largest city points in the same dataset to display them three times. Just assign them a minZoom value of "0" and make sure to call them in the code with "less than" in order to capture them at all the needed zoom levels.

Incremental Styling

In digital interactive maps, the goal is usually one of the following: to maintain a consistent look and feel across all zoom levels or to achieve a more or less generalized look and feel depending on the zoom level. To go about doing either of those things, the various parameters that are associated with the styling of features need to change incrementally by zoom level. These parameters include, but are not limited to, font size, line width, color, halo size, opacity, brightness, and point size. It would be great if there were a way to accomplish this programmatically with a couple of lines of code where one sets an initial parameter specification (e.g., "[zoom = 0] {line-width: 1.0;}") in the first line, and the next line tells the renderer to propagate that same style across all the rest of the zoom levels in accordance with the scale. Unfortunately, this isn't a common option at the time of this writing.[2] To autoadjust numerical parameters linearly, one technique is to multiply an initial parameter specification by a certain number for each zoom level like this:

```
[zoom = 0] {line-width: 1.0;}
[zoom = 1] {line-width: 1.0 x 4;}
[zoom = 2] {line-width: 1.0 x 16;}
```

The multiple, in this case four, is called a *scaling factor*. The way the tiling scheme works in standard web mapping, the mapped area quadruples at each higher zoom level. Thus, in the example above, it seems intuitive to have used a multiple of four. However, the example shows how a multiple of four quickly gets to be much too high a number (much too wide of a line width in the example). Adjusting the scaling factor downward through experimentation is one solution. A scaling factor of 1.23 has been shown to be more appropriate for some datasets.[3] In practice, digital map designers often set these parameters as constants and don't use scaling factors, though that doesn't mean that scaling factors are a bad approach.

Instead of setting a different parameter specification for each and every zoom level, it is often the case that *groups* of zoom levels are given a single specification. In our road line-width example, the map designer might want to display roads at zooms 5 through 8 at a line width of 1.0 instead of supplying separate line widths for zooms 5, 6, 7, and 8. Also keep in mind that size changes might need to be created for some zoom levels and not for others. For example, city labels might be coded to be six pixels at zoom 5, seven pixels at zoom 6, eight pixels at zoom 7, nine pixels at zoom 8, ten pixels at zoom 9, and 12 pixels at all higher zooms. At the lowest zoom levels, font sizes in particular need to be a certain minimum size in order to be legible, and at the highest zoom levels font sizes should remain at a certain maximum size as it is not necessary to create huge labels even when zoomed quite far in.

Being aware of pixel size at certain zoom levels is a good tactic as well. If the mapmaker has a very large point dataset with the potential to have tens or hundreds of points in each pixel at low zoom levels, the mapmaker might decide to resample the dataset so that there is only one point per pixel. However, this needs to be tested on varying output devices and resampling may not be ideal, in any case.[4] Color saturation is another variable that has special considerations. If the color saturation is increased by a scaling factor, the higher zoom levels may wind up with a more pronounced glow effect than intended. This, too, needs to be adjusted experimentally during the map design process.

Incremental styling by zoom level also involves generalization. *Generalization* in this context means exactly the same thing that it means for paper maps, except that the associated decisions must be made for each and every zoom level. For example, a map with road data could display all road types from highways to primary roads to local roads at zoom levels 10 and higher. This same map would get very cluttered at zooms less than 10, even if the road line widths were decreased. To solve this problem, most digital maps drop the local roads and even the primary roads as the zoom levels decrease. In our example, zoom levels 5 through 9 might show just the major highways, and zoom levels less than 5 no roads at all. This use of selected portions of a dataset for different zoom levels is a form of map generalization. It should be noted that in some cases where advanced cartographic technique is warranted, a dataset of locationally selective generalization might be used. For roads, this would involve decreasing the types of roads shown at lower zoom levels, but by different amounts depending on whether or not those roads are in rural or city locations, in order to preserve a visual density contrast between rural and city locations. This is something that would need to be done algorithmically.

Another, similar, generalization technique is to display certain features not by type, as in the previous example, but by size. Previously we discussed city population sizes, but generalization can also be enacted on polygon datasets via an area field or column. For example, a map of protected areas could display only the largest protected areas at the lower zoom levels and incrementally add in smaller and smaller protected areas as the user zooms in. Perhaps you only want the major national parks to be visible at zooms 5 through 9, but then want the minor parks to begin to appear at zoom 10. This is most efficiently carried out by providing the necessary area constraints and keyword assignments at the data level, by preprocessing it or by dynamically querying it using a WHERE clause if your data is stored in a relational database. While it's not an appropriate method for a thematic map showing the totality of the world's protected areas (since not all protected areas would be shown on a world scale), it is a great technique for maps that are meant more for navigational purposes or for maps meant to be basemaps.

Another form of generalization is the resampling or reduction of nodes, which results in smoother lines and polygon boundaries. A dataset that has been generalized is usually stored as a separate dataset from the original, though simplifying algorithms could be built straight into the map code.

Whichever way, the generalized data would be used at lower zooms and the nongeneralized data would be used at higher zooms. For example, you wouldn't want to have generalized state boundaries displayed at zooms 13 or higher as they will start to take on a cartoonish look. Conversely, you wouldn't want detailed state boundaries displayed at zooms 5 or less as they can appear too thick due to the large number of nodes displayed in such a small map area. Another potential pitfall of using generalized data at higher zoom levels is that the data will not line up with the other data on the map. If a state boundary doesn't line up with a more detailed park boundary, which your users know to be coincident in real life, the users will *not* think "Oh, the mapmaker must have just used a generalized state boundary instead of the higher-resolution one here." No, they will undoubtedly be thinking something more along the lines of, "This map is wrong! I can't trust anything on this map because it is wrong!" Remember, map users don't know much about how maps are made or that there are different datasets for different scales, so it's the job of the mapmaker to make the data line up right.

Any images used in the final map—such as icons (sometimes called *markers*) to represent points of interest or road shields—should also be incrementally scaled as the user zooms in and out. A single icon is not suitable for all zoom levels. At least three sizes of icons (small, medium, and large) are recommended as a starting point. In some instances, it is entirely appropriate to have a slightly larger icon for each and every zoom level, calling each individually for each zoom level. A trailhead marker should increase in size a small amount at each zoom level up to a certain point where it is appropriate for it to remain the same size. Icons that have labels in them, however, can get more complicated. Road shields, in particular, need to accommodate both the different lengths of the road names (often specified in a name length field in the database) but also need to accommodate the changing size of the road name labels at different zooms. A lot to think about!

Dot density mapping also presents challenges with zoom-level styling since a cluster of dots at a low zoom level may need to appear as separate individual dots at higher zoom levels. One might want to show circles around dense areas labeled with the number of dots within that circle, but then switch to showing the actual dots at higher zoom levels. One might want to show more of a heat map type of visualization for low zoom levels and more of an individual dot situation for higher zoom levels. Or, with extremely dense datasets, it makes sense to simply decrease the cluster density incrementally as the user zooms in. Whatever the strategy, the key to making this type of map work is in knowing the particular nuances of the dataset being mapped: how many points at high zoom levels are there in a single representative tile, and where are the high- and low-density locations? From this information, the mapmaker can place zoom-level restrictions (i.e., not render the highest or lowest zoom levels as necessary) and decide on whether or not to programmatically change the clustering by zoom level or to use separate datasets or separate visualizations by zoom level.

In the future, we may have automated ways to increase or decrease symbol sizes by zoom level, simplify geometries by zoom level, increase or decrease icon sizes by zoom level, and cluster or de-cluster by zoom level. The mapmaker might only need to supply a single symbol size, a detailed dataset, a single icon, or one point-density parameter and the software would extrapolate from that single reference to make the map appear perfectly at all zoom levels. Until then, it will be the cartographer's job to tweak and fiddle, simplify and enlarge, and reprogram as needed. And until then, we will have many different interpretations of what sizes appear best at which zoom levels and for what purposes. Digital maps, then, are far from being the end to artistic interpretation in mapmaking; they actually take the application of complicated symbology rules to a whole new level.

Repetition

Dealing with repetition in map code is less than straightforward. Grouping the code for several zoom levels at a time is the most obvious thing to do when they all have exactly the same symbology parameters. And indeed, that's the way to go if making sure the code is as short as possible is the goal. This method also helps to immediately indicate to others who are tweaking the code later that the grouped zoom levels have the same symbolization parameters for that dataset at those zooms. For example, in the code you might have something like this:

```
[zoom> = 3][zoom< = 5] {opacity: 0.5; point-size: 2.0;}
[zoom> = 6][zoom< = 8] {opacity: 0.4; point-size: 5.0;}
[zoom> = 9][zoom< = 11] {opacity: 0.3; point-size: 6.0;}
```

However, some mapmakers prefer to repeat the code for all the zoom levels, one by one, even though it means duplicating the parameters. The previous example, if set up with duplicated parameters, would look like this:

```
[zoom = 3] {opacity: 0.5; point-size: 2.0;}
[zoom = 4] {opacity: 0.5; point-size: 2.0;}
[zoom = 5] {opacity: 0.5; point-size: 2.0;}
[zoom = 6] {opacity: 0.4; point-size: 5.0;}
[zoom = 7] {opacity: 0.4; point-size: 5.0;}
[zoom = 8] {opacity: 0.4; point-size: 5.0;}
[zoom = 9] {opacity: 0.3; point-size: 6.0;}
[zoom = 10] {opacity: 0.3; point-size: 6.0;}
[zoom = 11] {opacity: 0.3; point-size: 6.0;}
```

The argument for this practice of duplication is that setting it up like this at the beginning provides an easy way of modifying parameters as needed during the map testing phase. For example, while it may have initially seemed like a good idea to assign a point size of 6.0 to zoom levels 9 through 11, perhaps testing indicates that a slightly larger point size is needed for zoom level 11. It is much easier to quickly change the 6.0 to a 6.5 for zoom level 11 for testing purposes when the code is initially laid out in this manner. Whether or not the map designer goes back and simplifies the code by grouping levels together after testing is complete may depend on whether the code will be shared (an argument for reducing the redundancy) and whether it is likely to be modified again at some point in the future (an argument for keeping the redundancy).

Testing a Multizoom Design

Testing of a digital, interactive map is all about making sure the right features are showing up in the right areas, with the correct symbology, in all locations, and at all zoom levels where they are supposed to be. To be clear, testing of the underlying data and user interface components is also an absolute must if the datasets are not from reliable sources with known accuracies, and if the application contains a complex interface design. However, here we'll focus on testing in the sense of "does the map look the way I thought I coded it to look?" Usually, testing will come in two phases: (1) the experimental phase, while coding is being carried out and (2) the rendering phase, when tilesets are being produced. Whether you are looking at a map in progress or a finished tileset that may need to be re-rendered, you'll be facing the same common issues. There are a whole host of things that can go wrong. The things to watch out for during testing are discussed in the paragraphs that follow.

Ensure that the data maintains topology (i.e., is coincident) at the highest zoom levels. To do this, carefully examine all the features at the highest zoom levels in various parts of the world, by eye, to make sure that things like state and country boundaries line up where they should, so that the shared lines have the same geometry and there are no dangling or foreshortened lines at intersections. Also examine where land and water features meet, so that rivers that separate political jurisdictions have coincident geometry with the political boundaries and so that ocean boundaries don't awkwardly cross river mouths (this can be apparent if the ocean color is different from the river color). If noncoincident geometry is discovered, it is highly recommended that the datasets be swapped out for datasets that work well together. Barring that, there may be a way to change the symbology of the more dominant layer so that it obscures the nonconformity, though this is a less than ideal fix.

Examine the layer order in the finished map. Are the features in the proper order in terms of layering? For example, airport runways might accidentally be hidden underneath airport perimeter areas (i.e., aerodromes). Or, road labels might be getting cut off where buildings overlap them. All kinds of things can affect the rendered order of the features, from the order in which the datasets are referred to in the code, to the order that they appear in associated configuration files. Once a fix is made, the other layer orders need to be rechecked to make sure nothing else broke in the process.

Seams between tiles can cause some unexpected effects when it comes to labels. Default minimum tile padding is generally built into the software so that labels don't get cut off at tile edges, but occasionally a label will be longer than the minimum padding, and the parameters will have to be manually adjusted. Panning around the map at each zoom level to examine tile seams is the usual method for checking for this problem. This is a less than ideal solution for enterprise-level maps, which may need to rely on more sophisticated color and pixel algorithms to cycle through rendered pixels for missing color when compared to sample maps of known problem areas.

Labeling density deserves a close look during the testing phase. The most important part of this testing is to make sure that labels don't overlap each other in dense areas. The same methods for fixing that problem also apply to labels that overlap with other features: increase the minimum distance or switch the Allow-Overlap function to "false." If the problem is simply that too many labels are showing up for the map's purpose (or too few), change the number of features being shown by eliminating features or otherwise changing the initial database query.

Examine labels for visibility. Do all features that are supposed to have labels actually show labels? Are there any features that have geometry but no corresponding label or any labels that appear without their corresponding geometry? Sometimes a label won't appear if it is overlapping another label, even if the feature itself appears. Sometimes the coding can be incorrect such that label behavior is not as expected. Generally, this can be tested at the lower zooms and expected to behave similarly at higher zooms. Lower zooms should therefore be tested first.

Different devices show colors differently. Does the palette chosen for the map look as intended across all possible output devices? Even if the colors appear to be somewhat different, does the finished aesthetic still hold and does the color gradient (if any) still make sense? Computationally sophisticated testing might involve rendering to all devices and then measuring the contrast between all pairs of colors to ensure that it is the same or similar to the development device. It may also involve simulating color blindness on all output devices and subsequent contrast testing to see if the colors still contrast enough for color deficient individuals. Most digital mapping efforts though, can be tested via visual inspection of the rendered maps.

Halos can be problematic. The most frequent issue is that label halos are simply too wide for the labels to remain legible. If this happens, the fix is to simply decrease the halo size. Halos around labels may also unnecessarily lower the label density by increasing the chance that they will overlap. In most cases, the halo thickness can be decreased to ameliorate these issues.

Halos around features (sometimes called *edge glows* or *vignettes*), such as land–water boundaries for example, can overwhelm small features (such as islands) at low zoom levels. Testing of land–water boundary halos should include at least one area with small islands.

In thematic density maps, the visual story can change dramatically depending on the user's zoom level. Make sure to test all zoom levels for brightness and density of features to ensure they make sense from a data standpoint. Restrict the zoom levels and adjust the brightness as necessary.

Make a list or spreadsheet of feature type and zoom level. Use this documentation to test if all the features that are supposed to be visible at each zoom level actually are visible at each zoom level. Check them off as the testing progresses. In some software, problems could include opacities accidentally being set to zero, or sections of styling code accidentally being commented out.

Check for styling continuity between zoom levels. Does all numerically based styling change smoothly from one zoom level to the next? For example, make sure that roads don't suddenly go from thick in zoom 5 to thin in zoom 6 and back to thick again in zoom 7. Behavior is normally expected to be the same regardless of region (unless you've coded a regional density difference), so it is permissible to test just one location for potential issues with this, though it might take quite a bit of back-and-forth with the code to make sure that the code is written the way that the features are appearing.

The level of generalization should be looked at for all features. Generalization can include the number of nodes (i.e., twists and turns) for line and polygon features and the density of all types of features: point, line, and polygon. Ensure that features don't appear too abstract or too detailed at each zoom level. To fix features that are too abstract, more detailed data must be found or restrictions whereby the highest zoom levels are not allowed (rendered) must be enacted. To fix features that are too detailed, smoothing and simplifying algorithms can be used on the data or presimplified data can be located. In the case of common datasets such as country administrative boundaries, you will be better off looking for an official dataset than creating one of your own.

In terms of color blending, where the mapmaker specifies different opacities for overlapping features, maintaining the right colors over all zoom levels and all potential combinations of features is difficult. Be sure to check the different combinations by identifying the maximum and minimum color blending situations and testing those in the output for aesthetics. These situations can include places where all opaque features overlap (maximum) or where only one is visible (minimum), as well as places where all colors overlap (maximum) or where only one color is visible (minimum). Gradients are particularly hard, if not impossible, to achieve well in a color blending situation, so be sure to test and make sure those colors combine to form the right hue.

You can see that there are a lot of things that can go wrong in a digital map project. It's best to pick at least one area that is well known to you to begin your testing. If that area looks correct, then look at (at least) one highly populated area and one rural area, one coast, one island, and one inland area. In other words, choose your testing sites with variety in mind. If the project is a complex one with many zoom levels and more than 10 or 15 datasets, make sure you enlist the help of a quality assurance/quality control (QA/QC) team to formally test the output if a professional, enterprise-level product is the goal. Formal testing is greatly enhanced if an issue-tracking tool and a central code repository are employed, for testing and fixing by one or more cartographers.

Endnotes

1. A few ideas for picking up development skills are: learn Python at http://www.learnpython.org (accessed October 31, 2013); take Harvard's free online course in computer science called CS50 at https://www.cs50.net/ (accessed October 31, 2013); watch Julie Powell's ArcGIS Viewer for Flex tutorials at http://www.youtube.com/watch?v=25uC2e3LsLk (accessed October 31, 2013); and use TileMill and MapBox with a free account to learn CartoCSS.
2. GeoServer is one technology that does have some autoadjustment capability for buffers and other situations.
3. Mapping Millions of Dots by Eric Fischer, July 22, 2013. http://www.mapbox.com/blog/mapping-millions-of-dots/ (accessed September 27, 2013).
4. A pixel size chart, with some assumptions. http://resources.arcgis.com/en/help/arcgisonline-content/index.html#//011q00000002000000 (accessed September 27, 2013).

Resources

Mapbox's documentation on styling for zoom levels outlines more ideas on how to turn layers on and off by zoom level and other important points. http://www.mapbox.com/tilemill/docs/guides/advanced-map-design/.

ScaleMaster is a table of map scales and common datasets suitable for each scale and is available (as a PDF file) from Cynthia Brewer. http://www.personal.psu.edu/cab38/ScaleMaster/.

Looking at how existing multiscale maps are created is a great way to get started. Take a look at the publically available code for **OpenStreetMap**'s "Bright" world map. Start by looking at the base-map style sheet here: https://github.com/mapbox/osm-bright/blob/master/osm-bright/base.mss.

Study Questions

1. What is a scaling factor and what are the benefits and drawbacks of using one?
2. Give an example of generalizing a dataset by size. Choose a geographic data type and produce a plausible chart or pseudocode snippet with appropriate area constraints by zoom level.
3. What are the problems associated with using a low-resolution dataset at a high zoom level?
4. Pick a major highway, such as I-25, find an appropriate road shield vector file for it (try Wikimedia), and create three shield sizes: small, medium, and large. Create the new shields in SVG format.
5. What is a map stack?
6. Discuss the Web Mercator projection. What are its strengths and weaknesses?
7. Pick an existing, publicly available digital dataset. Briefly describe the dataset and the map scales for which it is appropriate. Use Figure 9.5 in this chapter to indicate the zoom levels for which it is appropriate.
8. How is a zoom field used?
9. What can go wrong with halos on a digital map?
10. Does a map at a particular zoom level look the same in terms of scale on all output devices? Why or why not?

Exercises

1. Create a simple digital map of zoom levels 8 through 12 with only country boundaries (admin 0). Create a separate set of rules for the country boundary styles for zoom levels 8 through 10 and for zoom levels 11 through 12. The style rules should include color and width.
2. Create a simple map of zoom levels 5 through 15 with only country boundaries (admin 0). Use generalized administrative boundaries for countries at zoom levels 5 through 8. Swap in a more detailed administrative boundary dataset for zoom levels 9 through 15.

Appendix A: Layout Sketches

A printed layout can be put together in a million different ways. The examples that you find here are specific to the type of printed layout that contains margin information in an actual margin (i.e., set apart from the map element) rather than floating over the map element. They are drawn in the same format as recommended in the emphasis map section of Chapter 3, "Layout Design." These sketches are meant to provide examples of good layout technique and to give inspiration on how best to organize multiple elements on a layout. They are obviously not comprehensive of all the ways these elements can be arranged.

There are several things to point out here: first, you will notice that all the sketches are formatted to look somewhat like real sketches on paper. While the original intent was to provide actual hand-drawn sketches in this appendix, it turns out that hand-drawn sketches are not easily printable in book form due to their naturally varying line widths. As a solution to the problem, the sketches are presented here in computerized form, but this does not mean that your layout design attempts need to be done the same way. Indeed, it is not even advisable to do so (unless perhaps your map will be published in a printed book, of course).

Why should your emphasis maps be in sketch form? Because sometimes we just need relief from computerized constructs. Switching from your everyday environment (the computer) to a pen and paper can be liberating to those creative processes that are just waiting to come out of latency and get to work for you. Another reason to do this with pen and paper is because it is much faster to create a layout sketch on paper than it is in a graphics program, even for those of us who are seasoned graphics professionals. Speed is important so that you can get as many ideas out as possible during this very first, highly creative stage. Later on, when you are ready to finalize your layout design, you could transfer it to a digital graphic if it suits you. Finally, the last reason to sketch out your layout design on paper is because you will likely create several at a time as your ideas evolve. The human eye can quickly and easily read and interpret your own handwriting and sketching much more easily than it can computerized fonts. With a handful of sketches to go through and from which to pick the best ideas, the handwritten ones will be easier to get through.

The layout ideas found in this appendix are meant to help you get some ideas on how to construct your own emphasis maps with your own variables. Again, they are examples of margin-type layouts. That is, the margin information is not found free-floating on top of the map element and is, instead, neatly contained within one or more boxes surrounding the main map element(s) (except the scale bar and north arrow, which can and many times should be placed directly on top of the map element). While my point of view is that this type of page layout is superior to the free-floating page layout and is generally an easy technique to utilize to make your map look more modern and professional, I am by no means trying to say that free-floating margin information is unacceptable. If you find that a free-floating design is the best thing for your particular needs, you are on your own in terms of where to put the margin information. Indeed, the reason I have not included such sketches here is that free-floating margin information layouts are dependent on the shape of the thing being mapped. So if the map element were of the African continent, for example, you might float the margin information in the lower-left quadrant of the page where the ocean is. Other map shapes would lend themselves to other possible margin-element groupings.

Thirdly, there are many different sizes of maps shown in these sketches. They are drawn to either a 1:1 or 1:2 scale representing layouts in letter-size, ANSI C, ANSI D, and ANSI E standard sizes, as noted in inches. They can be adapted to fit any size page, of course, especially given that these are supposed to represent the general shape and flow of your page layout rather than absolute sizes and placements. Most of the larger sketches were made with conference poster style in mind, though one of them is more of an up-close reference map design. The smaller, letter sizes, were made with up-close viewing in mind as I envisioned workgroups passing them around at meetings, or including them in a client report. These differences are noted in the caption for each sketch.

Line widths and relative darkness of the lines are representative of the amount of weight each item should have in the overall scheme. Thin gray lines denote areas where the viewer will look last, while thicker black lines indicate areas where the viewer will look first. For the most part, the boxes and lines are shown as a means of suggesting where things will go on the layout; they are not necessarily to be taken as literal representations of what should be on the final layout. In other words, you do not need to have a thick black line around the map element nor do you need thin gray boxes around inset map elements. Not all layout elements are shown, just enough to get the general idea of shape, form, symmetry, balance, and emphasis.

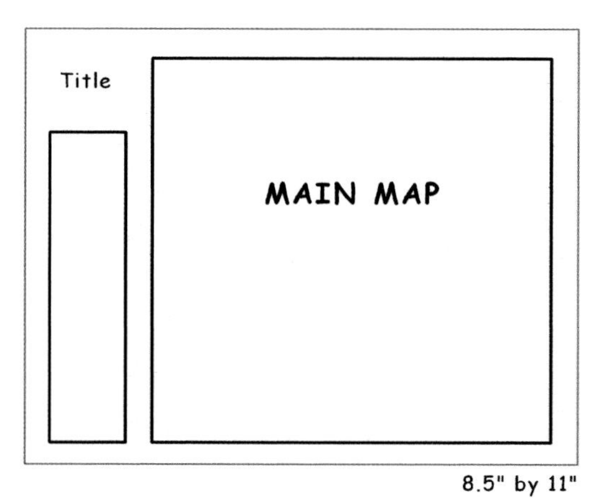

FIGURE A.1
This letter-sized layout is somewhat unique in that it has the margin elements organized on the left side of the page instead of the right side or bottom. If the viewer is expected to look first at the margin information and then at the main map, this could be an appropriate design. Alternatively, this design can provide some visual relief for cartographers and audience alike if the more typical design types have become hackneyed.

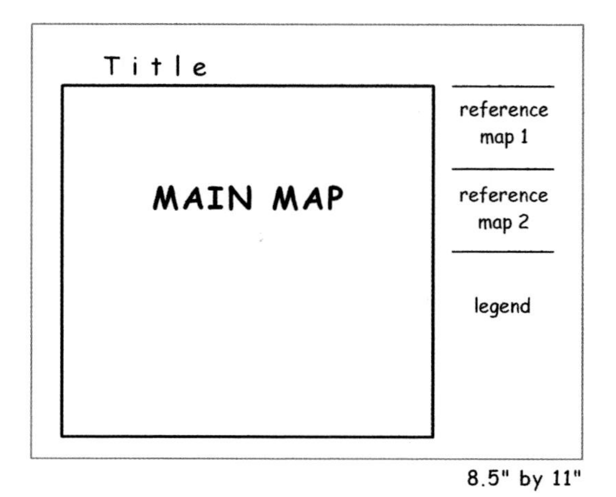

FIGURE A.2
I envisioned that this letter-sized layout would provide a good way of communicating large-scale information while still providing the larger spatial context for the information in the reference maps. A likely map candidate for this layout would be a zone map of a customer's property and its adjacent properties, with the reference maps illustrating the overall zoning for the region at varying scales.

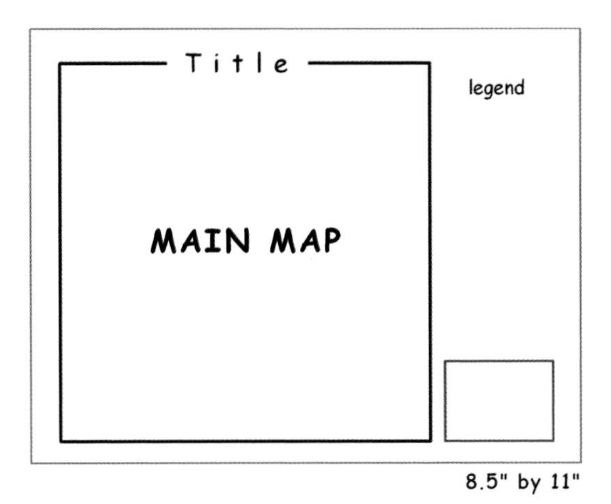

FIGURE A.3
While there won't be much room on this letter-sized sheet for a large map element, it is arranged so that a lengthy legend can be placed alongside the map. This is appropriate for large-scale maps that have a lot of colors and symbols to decipher such as soil or geology maps of individual properties. A small reference map can also be included to show where the main map is in relation to major landmarks.

FIGURE A.4
Putting the title off to one side, as on this letter-sized sheet, gives a layout a slightly more modern appearance than if it were located at the top, center of the page. Room is left on the bottom for a small amount of margin information.

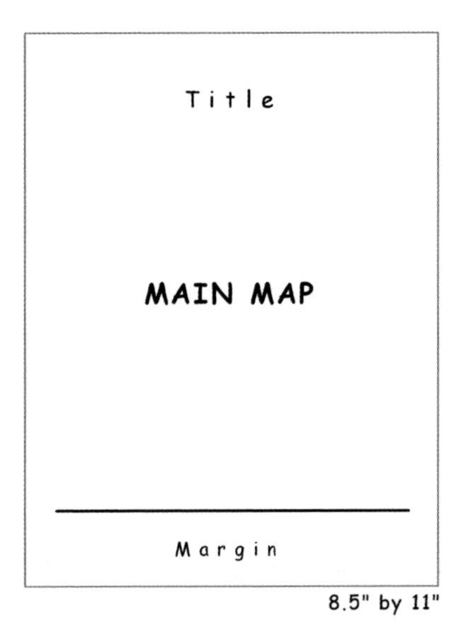

FIGURE A.5
This letter-sized sheet uses a simple line to separate the map from a small amount of margin information (the line is meant to be a literal line on the finished layout in this case). The title can float over the map if the map has white space to allow it. Alternatively, the title can be surrounded by a white halo or mask to differentiate it from the map element or it can be moved to the left or right until it is not obscured by the map.

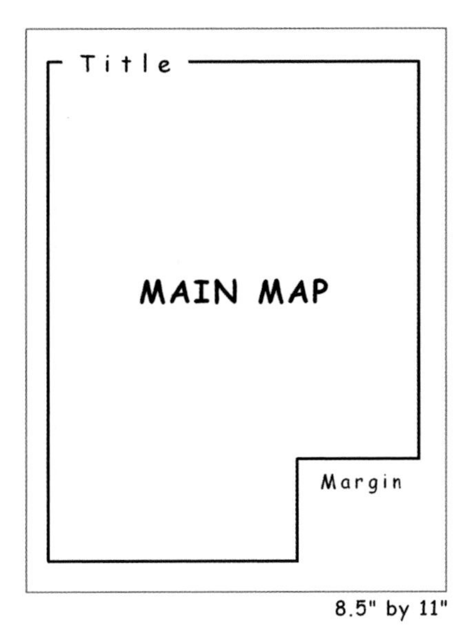

FIGURE A.6
This square margin box takes up a small part of the main map on this letter-sized sheet. It is positioned to allow most of the viewer's focus to be on the main map element and the title.

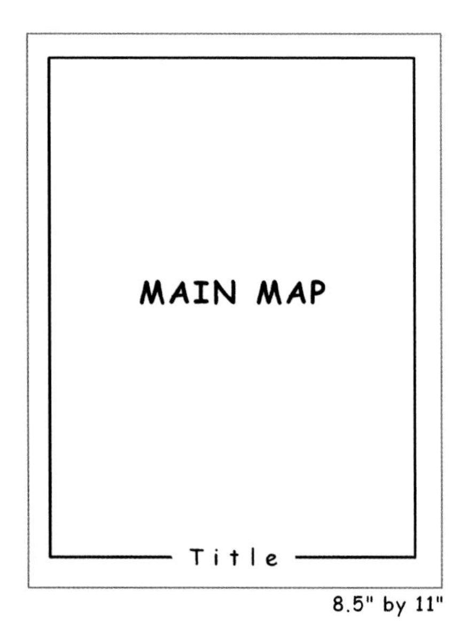

FIGURE A.7
Letter-sized sheets aren't always big enough to show your entire map element without sacrificing some margin information. This layout takes advantage of the entire page, squeezes in a title at the bottom, and perhaps includes a scale bar and north arrow on top of the main map element.

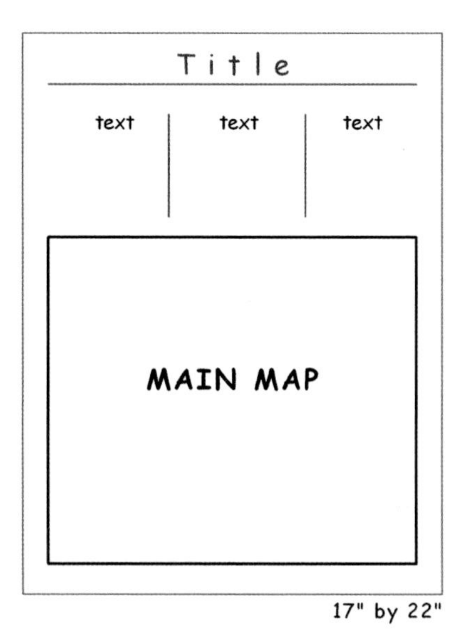

FIGURE A.8
This straightforward ANSI C-sized layout shows how you can arrange things so that the text is slightly more in focus than it usually is when relegated to the right-hand side or bottom of a page. In this layout the text is almost of equal importance to the main map element.

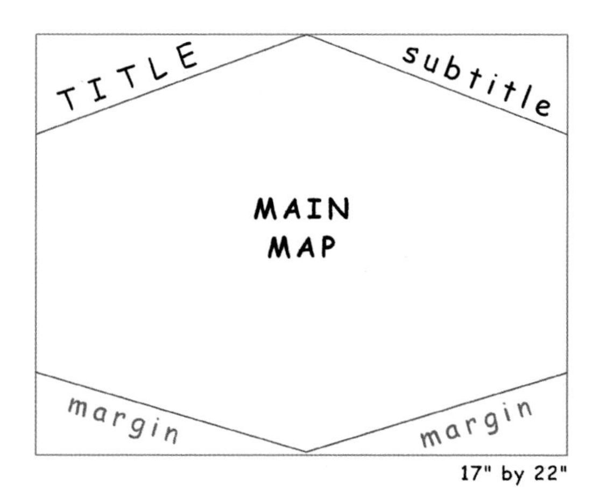

FIGURE A.9
This ANSI C-sized layout is definitely not your normal scientific-type layout, but could suit a variety of nonscientific mapping purposes where overall design is most important. It is particularly suited to small-scale maps like a world map or country map.

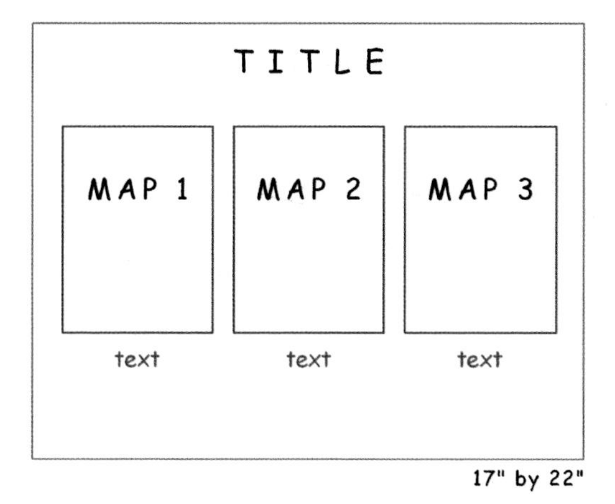

FIGURE A.10
When designing a layout with a series of maps (usually with the same extent but different time periods or different data), it can be helpful to lay them out side by side like this, aligned horizontally. Occasionally you'll see map-series layouts get overly fancy by staggering the maps and filling in the white space with other graphics. That approach takes away from the central point of the layout, which is to compare and contrast the maps. Putting them side by side, as shown on this ANSI C sheet, is a much better way to get the point across. Incidentally, it is usually best to leave the border boxes off of the maps (they are only shown here to suggest where the maps will be).

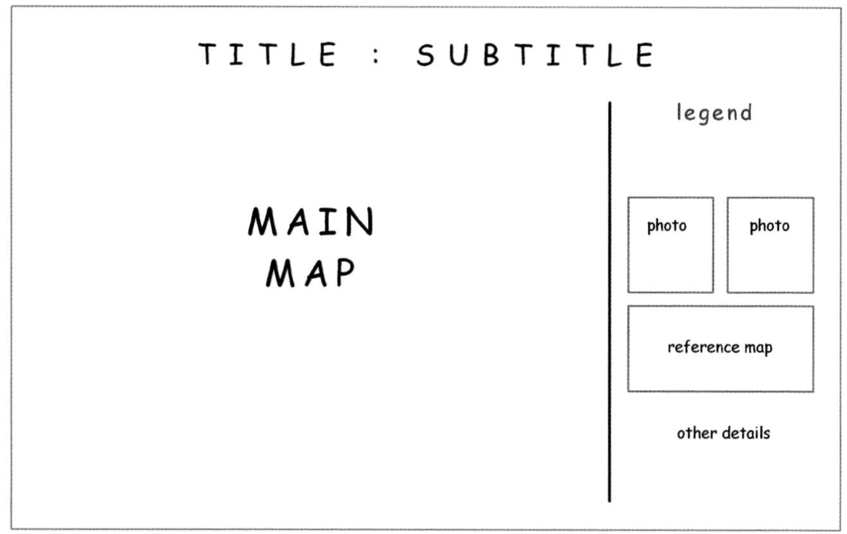

FIGURE A.11
ANSI D-sized sheets are long and narrow. They are ideally suited for times when you need a large amount of margin information, like in this layout where multiple photos, a reference map, a big legend, and plenty of space for other details take up the right-hand quarter of the page.

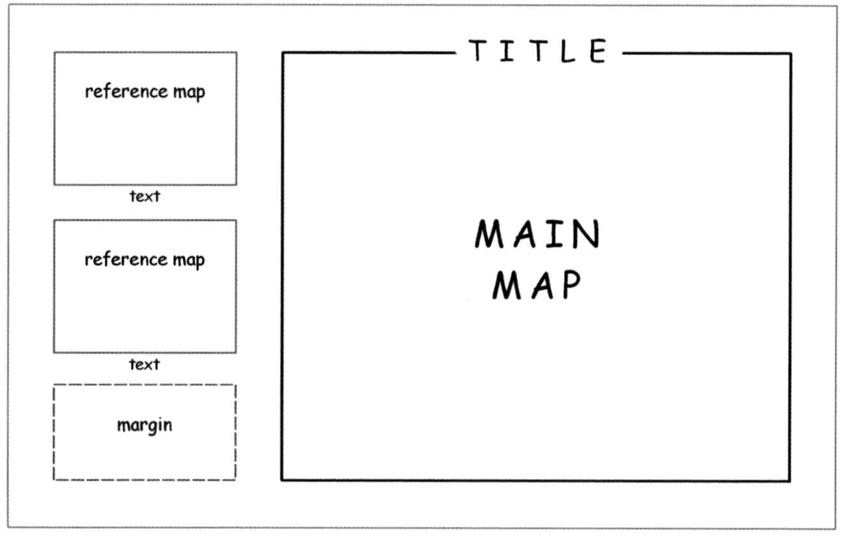

FIGURE A.12
Margins do not always have to be on the right-hand side or bottom of a page. In this case, I envisioned a map where the reference maps provide crucial detail for understanding the main map, and are therefore placed in the upper left where the eye may glance first. Each reference map is explained with some brief textual information directly below it on this ANSI D-sized sheet.

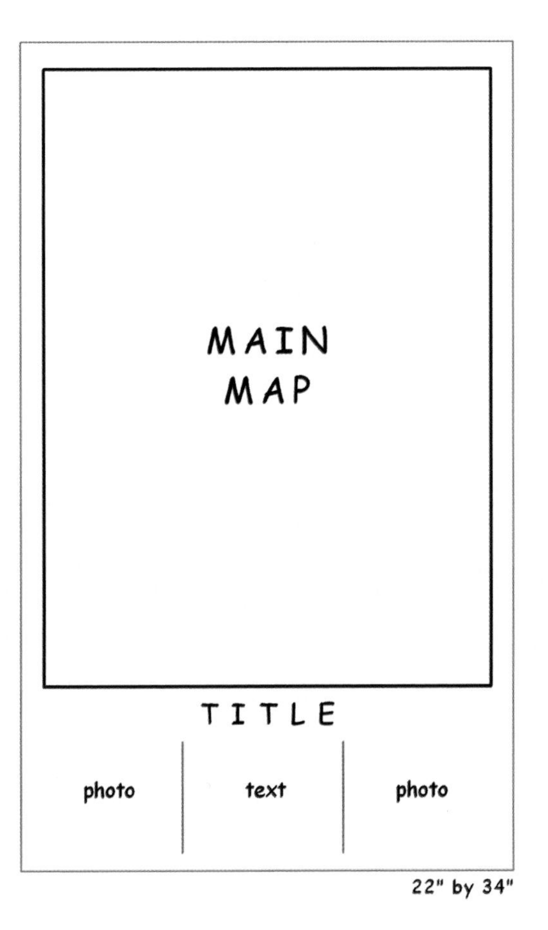

FIGURE A.13
Maps of longish study areas (California or Chile would be two examples) would work well on this ANSI D-sized page with plenty of room for a long narrow map element. With a couple of photos surrounding some textual information, and a title between, it achieves a balance while still allowing the main map to be the central focal point.

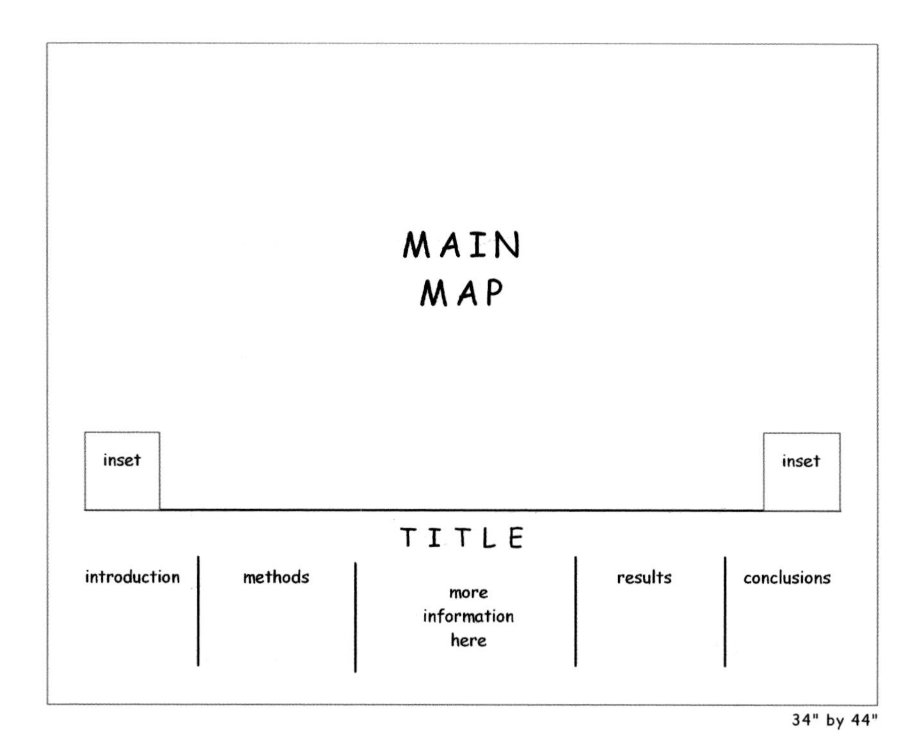

FIGURE A.14
This is a poster-style layout on ANSI E-sized paper. The idea here is to allow enough space to provide a huge eye-catching map while still retaining some space for very brief text concerning the study. This layout style takes advantage of the fact that most poster viewers will look at and read a poster from left to right by having the introduction, methods, results, and conclusions sections span the page in left to right order.

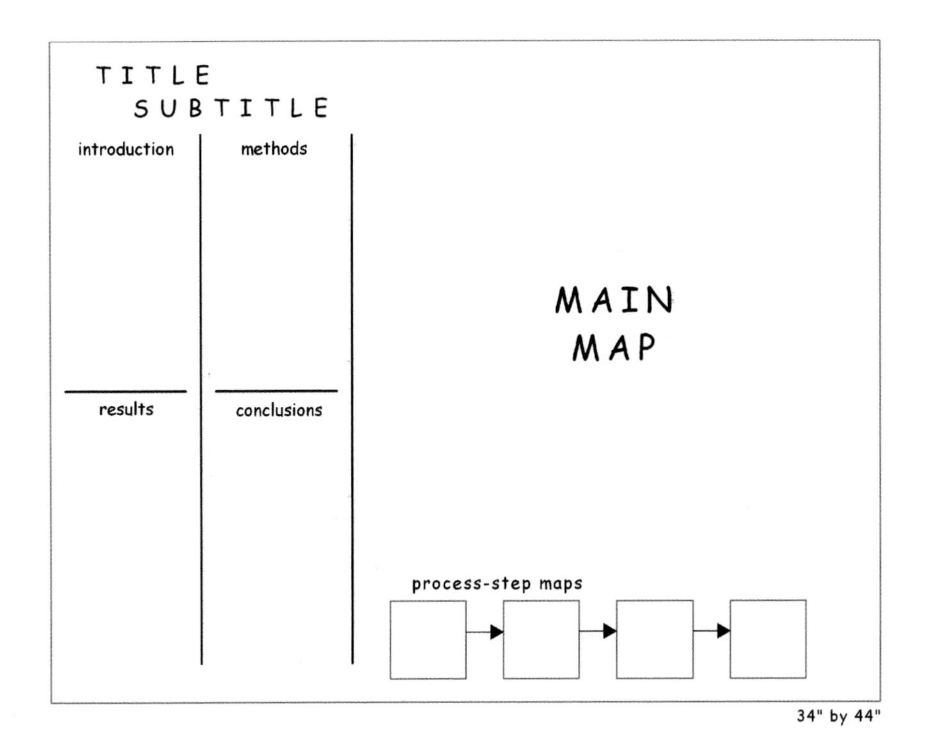

FIGURE A.15
Sometimes process-step maps can convey the methods used to derive the main map much more clearly and memorably than words can. This ANSI E-sized sheet employs this tactic as a means of graphically illustrating the study for a conference audience. The poster viewer could get a quick overview of the study methods via the small maps and look for more information in the text as needed.

FIGURE A.16
This ANSI E-sized layout leaves most of the available space to the map element for close-range viewing of detailed map data. I envisioned this as being part of a series of layouts for a large project, so there is a section at the bottom for the index map that will show which part of the overall map grid this map represents.

Appendix B: Map Examples

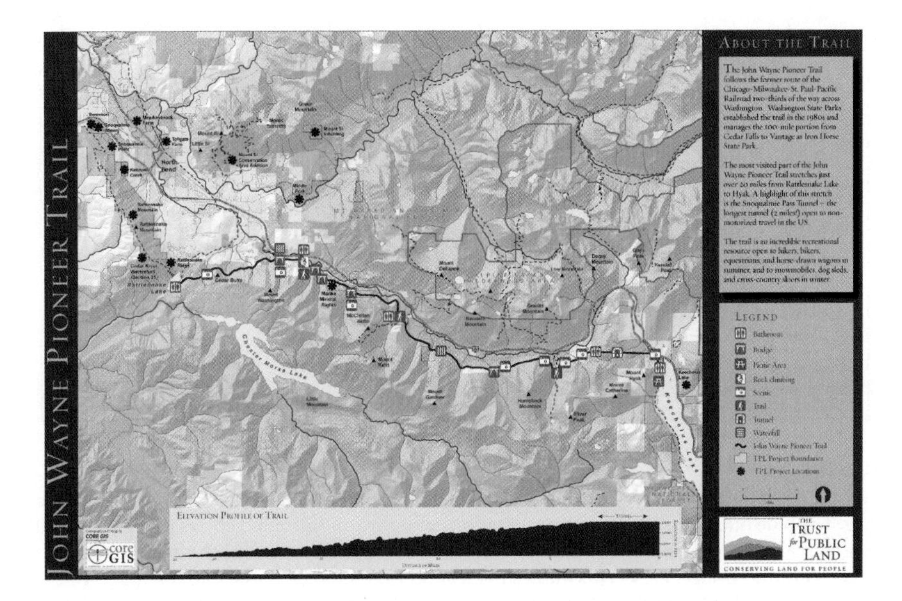

FIGURE B.1
This map is an outstanding example of an attractive, well-organized design that communicates effortlessly. With the use of well-organized margin elements, a generous amount of detail, well-placed labels, and standard symbols, visitors to the John Wayne Pioneer Trail will have no trouble locating themselves or the information and services they need. A dark background unifies the entire page while giving it a modern look. The two major titles are in a small-cap font that is lighter in color than the background, tying in well, as do the lighter-colored margin boxes on the side. Colors, fonts, alignment, and many other details were carefully and thoughtfully coordinated. (Designed by Matt Stevenson, Principal, CORE GIS, LLC for The Trust for Public Land. With permission.)

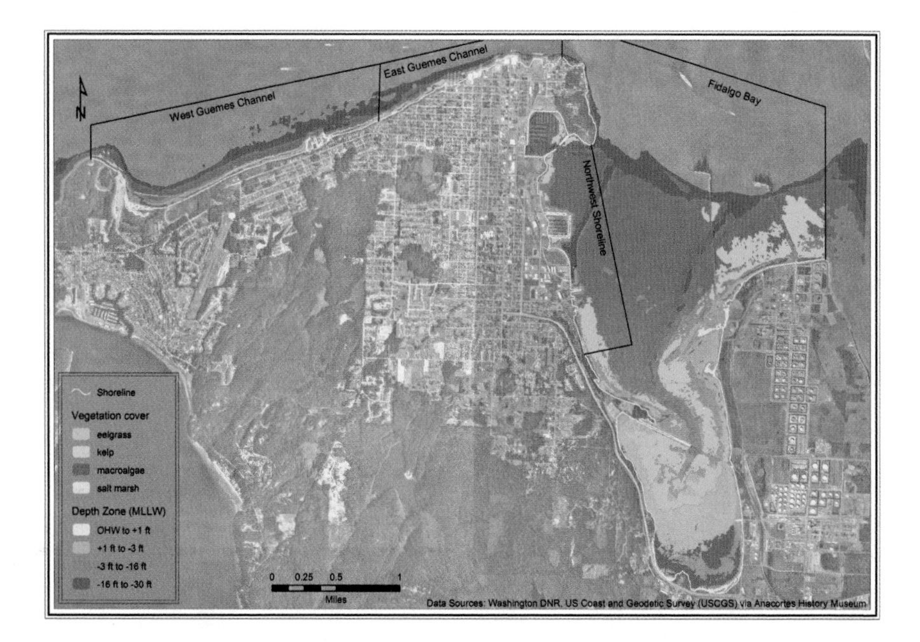

FIGURE B.2
This map of vegetation cover and depth uses black-and-white aerial imagery as its background layer. The blue hue of the bay and the vegetation colors are all saturated colors that go well with the grayscale background. Furthermore, the legend background matches the dominant color of the background image: gray. This allows the legend to remain secondary to the map, whereas a white background would have drawn too much attention to it. The data sources are neatly tucked into a corner. A small amount of white space between the map and the map's frame allows for a gradual transition to the white space of the report page on which it was placed. It also matches the small amount of white found in the image itself. (Designed by Allison Bailey, Principal, SoundGIS, for the Washington Department of Fish and Wildlife. With permission.)

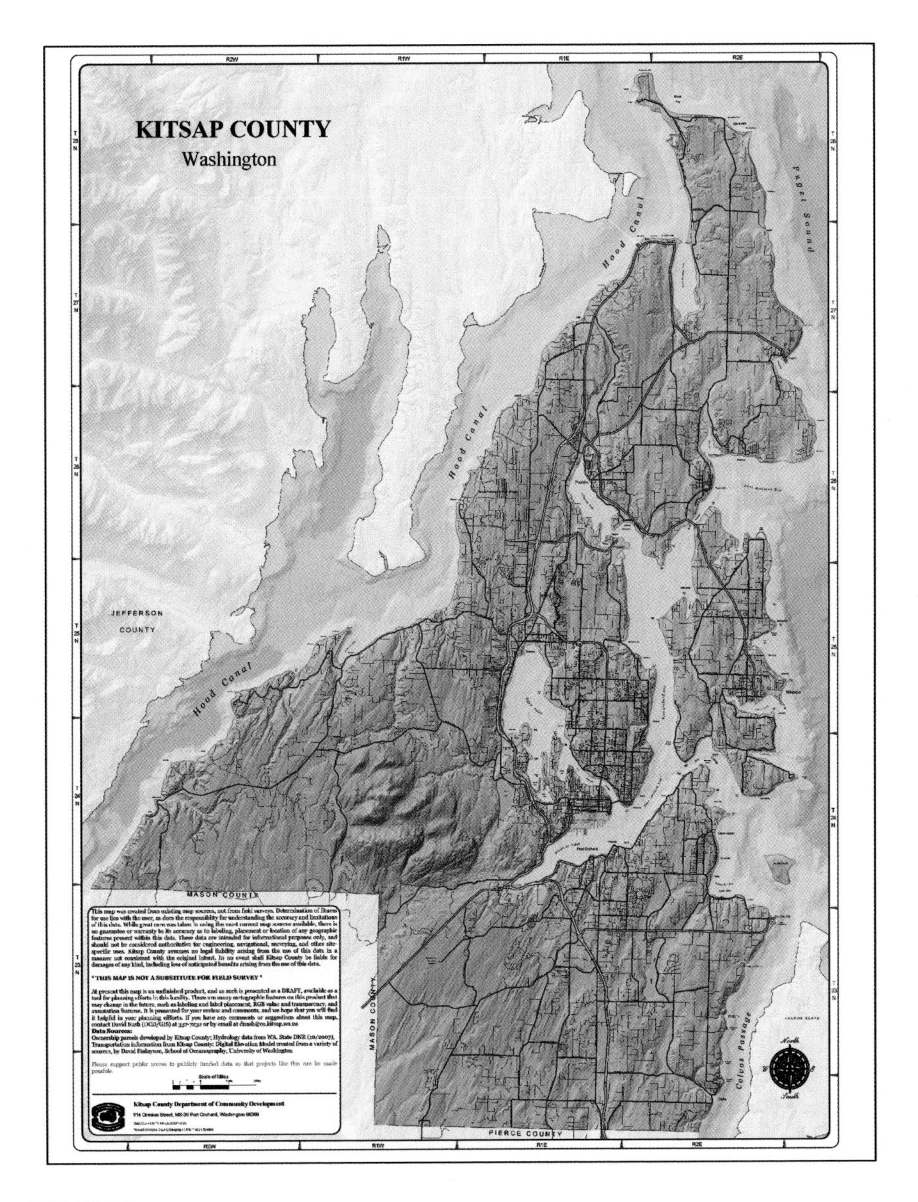

FIGURE B.3
This map serves as a basemap template for other Kitsap County government data. With its intuitive road symbol levels and hillshade with hypsometric tinting, it provides the spatial context for all other data. The darker tinting on the county focuses the viewer, while the lighter tinting on the surrounding areas is both elegant and unobtrusive. Almost the entire letter-sized layout is taken up by the map element to allow maximum use of space. The lengthy disclaimer is required by the department and is tucked well into a nonfocus corner of the layout. The title serves as a simple placeholder. Standard background symbols for roads, lakes, and elevation make a legend unnecessary. (Designed by David Nash, GIS Analyst, Kitsap County Department of Community Development. With permission.)

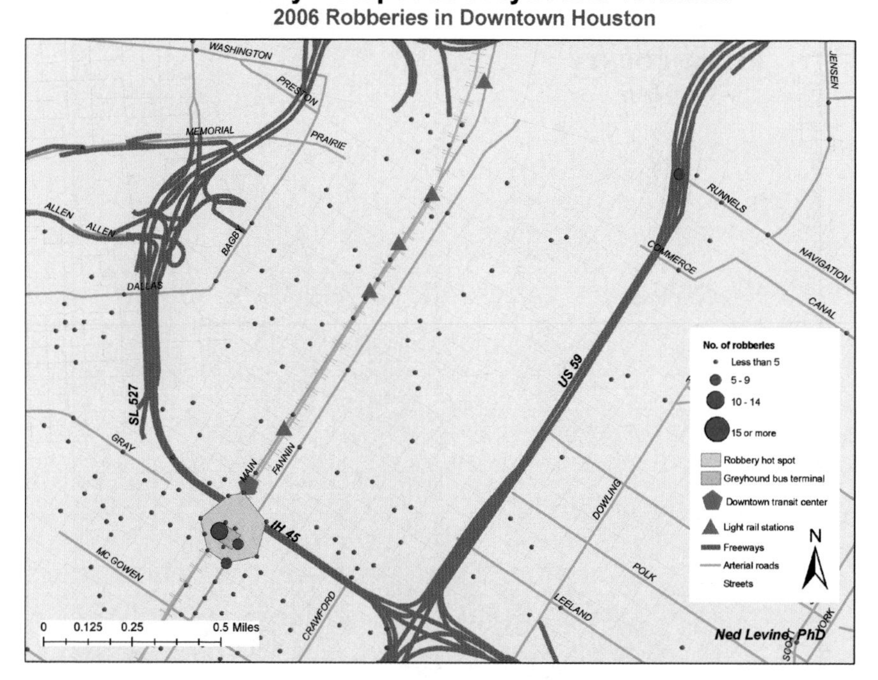

FIGURE B.4
This map illustrates two related statistical analyses: robbery density and robbery hotspots. It allows you to instantly verify the hotspot visually by comparing it to the large circles that indicate numerous robberies in that location. Some ancillary data help to further our understanding of the situation, including the location of a bus terminal that may be related. In case a viewer does not notice that connection, the map designer has clearly highlighted it in the map title. Road symbol levels and clear road labeling give good contextual background for data that are intimately tied with streets. (Designed by Ned Levine, PhD, Principal, Ned Levine & Associates. With permission.)

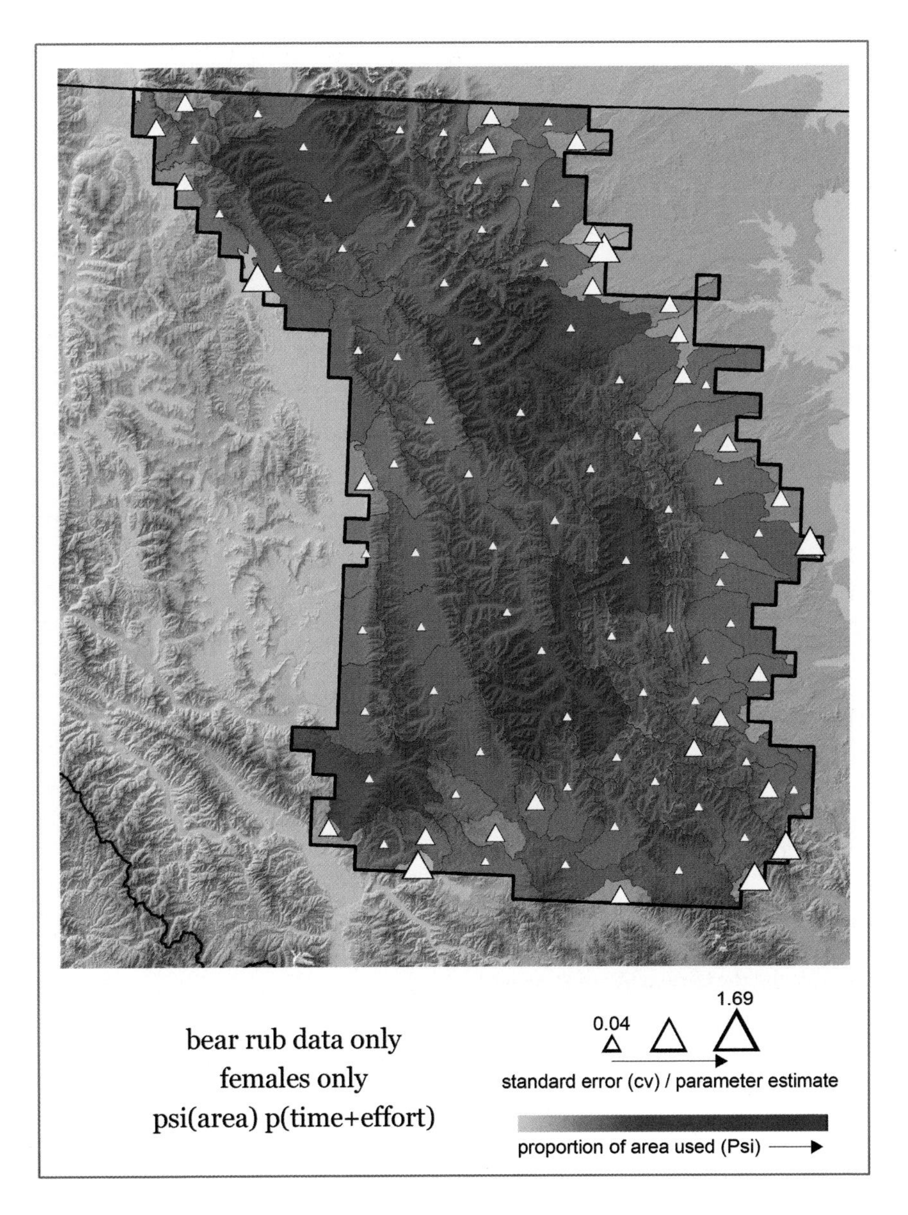

FIGURE B.5
This map, which shows the results of a simple model that estimates the proportion of area used by grizzly bears in the Northern Continental Divide Ecosystem, Montana, is part of a series of maps in a report. The other maps in the series share the same scale, location, drainage basin data, and elevation data, so those items are explained on a separate page. The three text lines in the margin explain the particulars of this map so the report reader can quickly see how this map is different from the others. The color ramp is used to highlight the number of bear rub markings found in each basin, which could have been from one very active bear or from several less active bears. (Designed by Jeff Stetz, Biologist, U.S. Geological Survey.)

FIGURE B.6
This vibrant map of Australia squeezes in quite a few labels while maintaining legibility and text hierarchy. Note the larger letter spacing and lighter gray of the Australian state and territory labels in contrast with the black city labels. Note also the splined text of the water features and the understated yet eminently usable scale bar. (Designed by Dan Bowles, *Australian Geographic* and the Cartographic Division.)

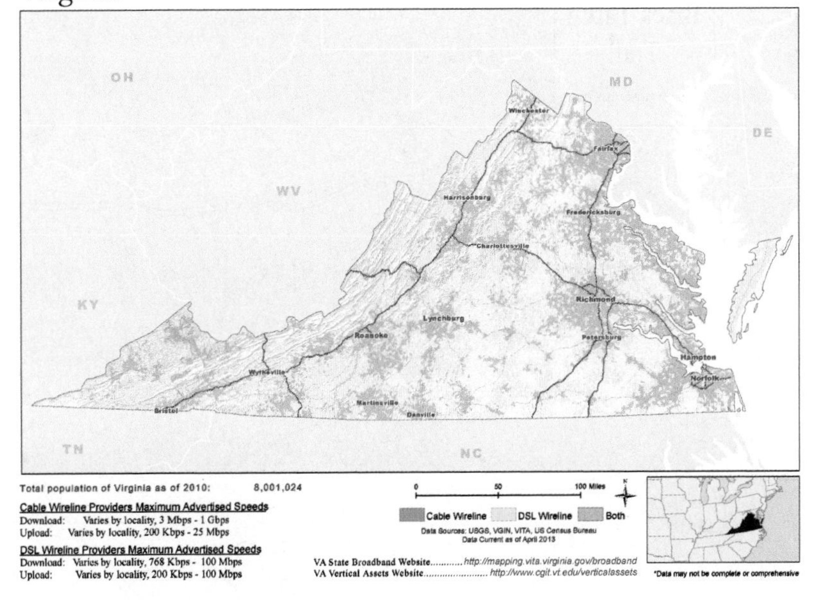

FIGURE B.7
This map is one of thousands in a series made at various scales (state, planning district, county) for the Virginia State Broadband Mapping Initiative. It shows cable wireline and DSL broadband Internet coverage in the state of Virginia as of April 2013. Some statistics are reported in the lower left portion of the map, while an easy-to-understand locator map appears in the lower right. Notice that the state names are in a large but light gray font. The city labels, in contrast, are much smaller, but in a black font. (Designed by Matt Layman, Virginia Tech Center for Geospatial Information Technology. Virginia Broadband Map Book Portal (2013). [VA_CableDSL, 2013]. Virginia Tech Center for Geospatial Information Technology. With permission.)

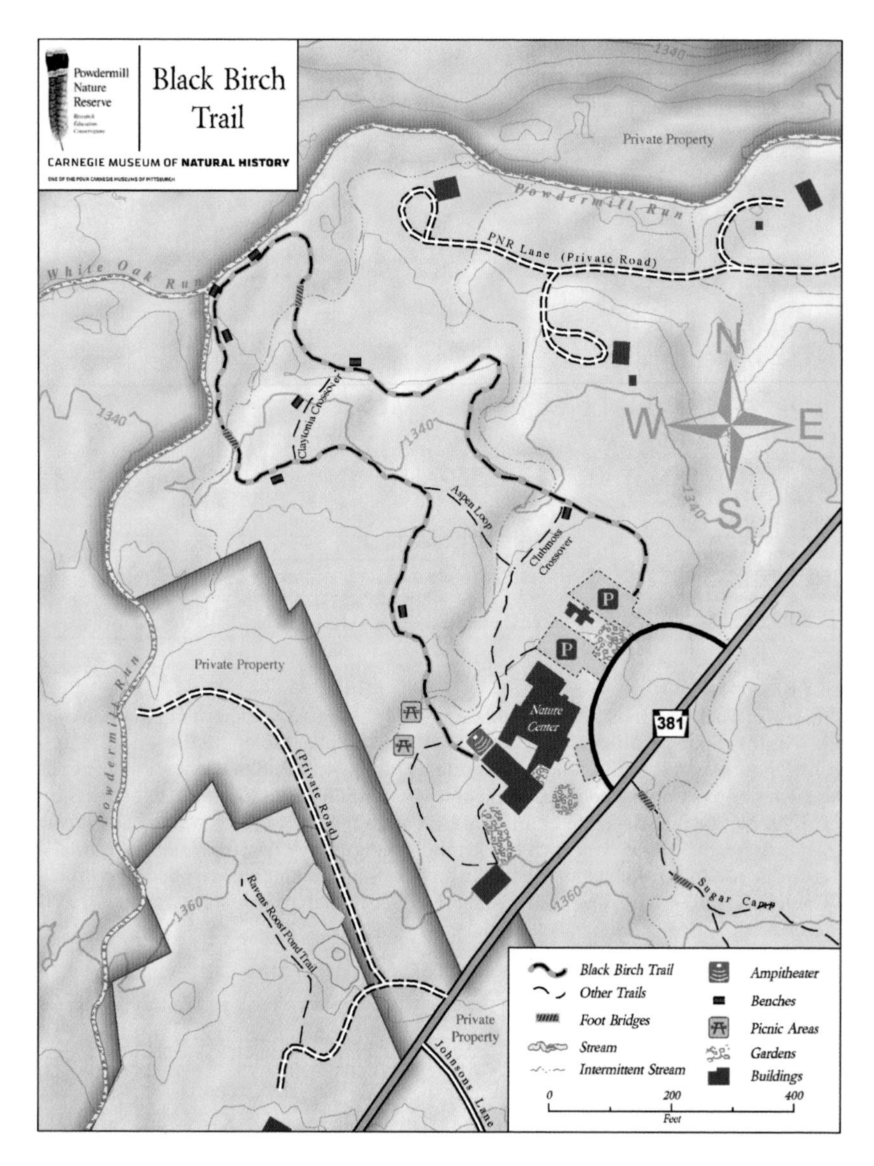

FIGURE B.8
The Black Birch trail is located on land owned by the Carnegie Museum of Natural History in Pittsburgh, Pennsylvania. A gray-shaded trail line with interior dashes highlights the trail well against the green background. Notice how the background has a subtle hillshade and hypsometric tint along with well-marked parking locations: all important elements for trail maps. (Designed by Michael Bowser and James Whitacre, Carnegie Museum of Natural History. With permission.)

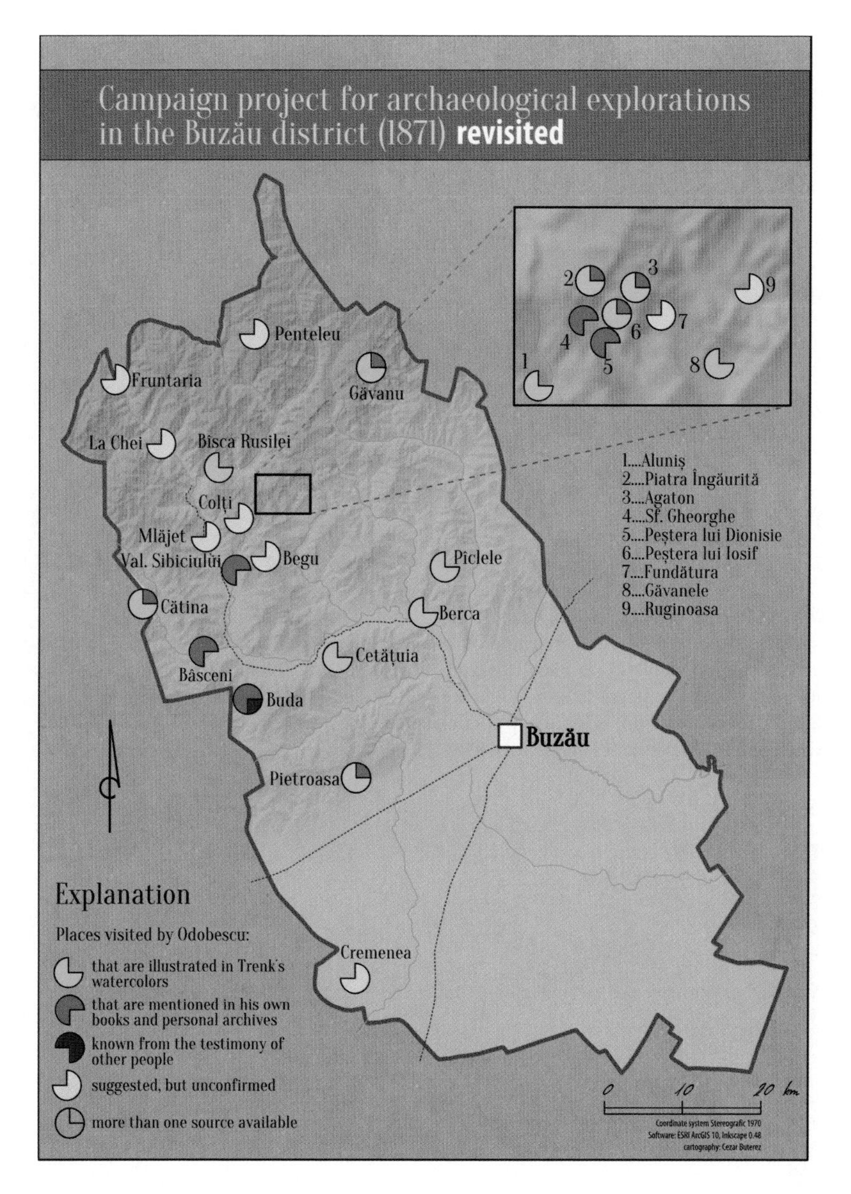

FIGURE B.9
This map focuses on points where a famous Romanian archaeologist, Alexandru Odobescu, led an exploration project in 1871. The map appeared in the *Mousaios* journal issued by the Buzău County Museum, next to an article about Odobescu's travels. The map title is clear, both in content and style. The district is clearly defined, while the surrounding empty space is kept interesting yet unobtrusive by the use of a complementary color palette. Inside the district, the topography is suggested but doesn't overwhelm the map, leaving the visited point locations to stand out. The inset map zooms in on a particularly data-rich area, and illustrates good use of leader lines. (Designed by Cezar Buterez, PhD Student in Human Geography, University of Bucharest. With permission.)

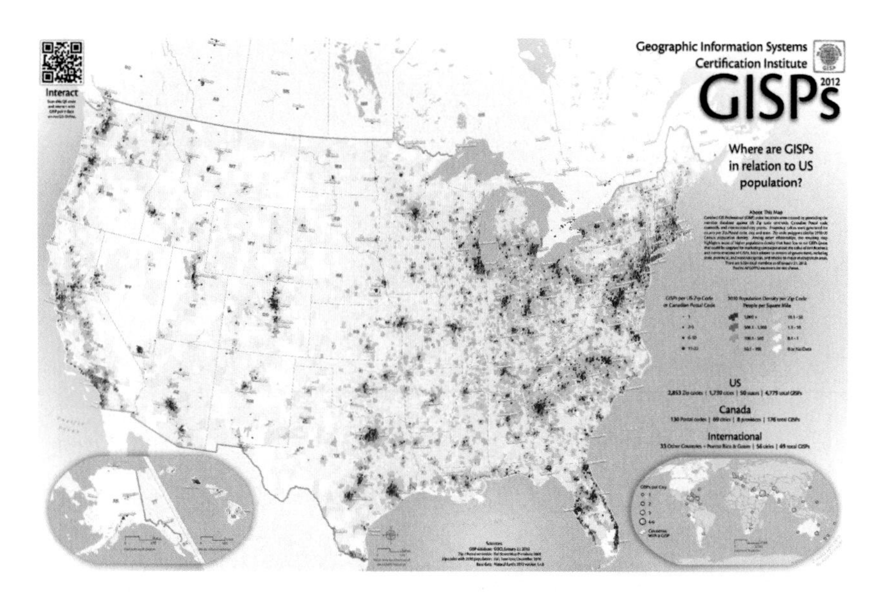

FIGURE B.10
This layout, meant for a large-format print, has well-organized and aligned margin elements. The map designer made good use of graphic design techniques, such as color fade-outs and shadows, to provide figure-ground definition. The central focus of the map is clear even though the layout contains a lot of supporting information. (Designed by Kyle Schaper for the GIS Certification Institute. With permission.)

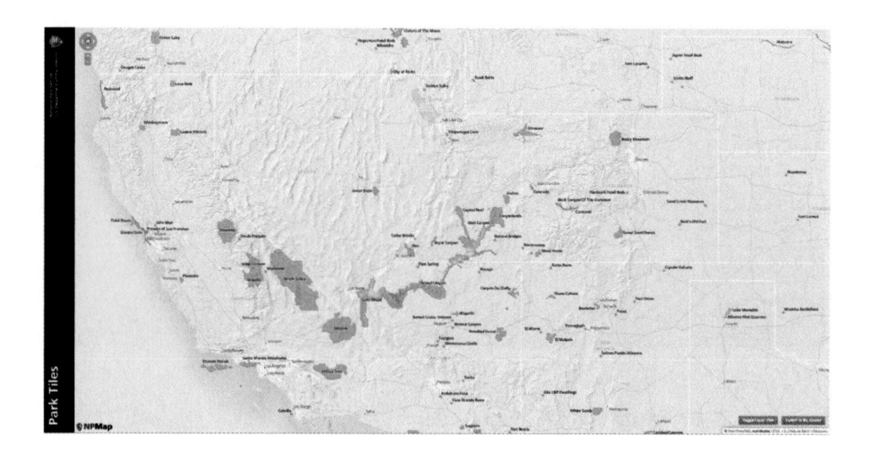

FIGURE B.11
Full-screen layouts for web maps are highly effective, as this screenshot of the National Park Service's Park Tiles shows. With only a left-hand title bar and a few buttons at the bottom-right, the map takes center stage, is easy to navigate, and promptly helps the viewer gain an understanding of where US parks are located. The hillshade basemap services the park concept well, while the much more subdued water areas allow the focus to remain on the land portions of the map. (Designed by Mamata Akella, National Park Service. With permission.)

Appendix C: Color Swatches

The color swatches in this appendix are the same swatches that appear in Chapter 6, "Features." Each section of that chapter has one or two swatches directly beneath the section title. They are repeated here to provide a handy reference to the RGB values for the colors. By no means are the palettes for these feature types to be confined to the colors found here. These are simply inspirational pallets that can be used as starting points for creating your own colors or can be used in their entirety as you deem appropriate. These palettes can also be used for other feature types that aren't described in the Chapter 6 or used to develop color schemes for entire layouts.

Roads	Rivers and Streams	Bodies of Water	Cities and Towns	Political Boundaries
255 255 255	157 159 166	237 235 245	255 255 0	255 255 255
225 225 225	67 154 184	224 228 204	0 0 0	137 112 68
255 0 0	0 111 158	167 219 216	0 255 0	255 99 0
255 170 0	0 52 69	115 178 255	50 50 50	52 52 52
255 255 0	5 9 130	126 190 163	137 112 68	104 104 104
115 76 0	190 255 232	53 50 138	255 255 255	230 64 33
78 78 78	99 222 183	83 160 142	255 0 0	176 186 207
0 0 0	169 177 194	22 0 199	178 178 178	222 170 102

Fuzzy Features		Elevation	Hillshade	Parcels	Currents
212 136 89		255 255 255	255 255 255	255 255 255	0 0 0
98 86 104		178 178 178	220 220 220	0 0 0	255 255 255
63 122 59		110 83 45	200 200 200	211 255 190	230 239 255
156 100 45		153 137 87	180 180 180	255 255 190	255 253 247
104 104 104		252 186 3	157 157 157	197 0 255	35 122 204
255 255 255	255 255 190	38 115 0	130 130 130	245 162 122	240 48 0
233 155 235	201 155 235	255 255 190	100 100 100	255 211 127	216 24 24
255 255 255	230 178 106	190 232 255	0 0 0	215 158 158	168 0 0

Wind	Temperature	Land use and Land cover	Trails	Utilities
0 0 0	180 0 0	235 109 105	0 0 0	76 0 115
255 255 255	228 59 36	206 166 138	100 100 100	100 100 100
230 239 255	201 124 0	255 224 174	255 255 255	255 190 190
255 253 247	233 76 19	166 213 158	255 167 127	0 112 255
247 236 210	244 253 8	117 181 220	255 0 0	76 230 0
50 50 50	230 239 255	188 219 232	215 194 158	255 85 0
130 130 130	198 195 215	200 200 200	158 73 3	255 255 0
232 190 255	38 123 172	189 221 209	117 29 0	150 163 0

Impervious Surface	Basins	Building Footprints	Soils	Geology
54 42 61	209 252 120	100 100 100	89 72 45	218 230 120
170 71 41	184 255 0	225 225 225	117 93 39	190 167 117
61 0 22	255 255 8	145 99 84	145 125 47	191 94 52
217 215 202	249 149 199	233 76 19	186 131 48	147 191 166
137 116 151	25 175 237	244 253 8	207 159 89	235 117 0
140 145 117	255 85 0	0 169 230	245 255 122	224 137 195
71 83 80	153 126 88	197 0 255	240 137 102	253 225 79
197 173 155	177 201 207	255 211 127	240 230 198	204 31 255

Index

F

Q

R

T

U

V